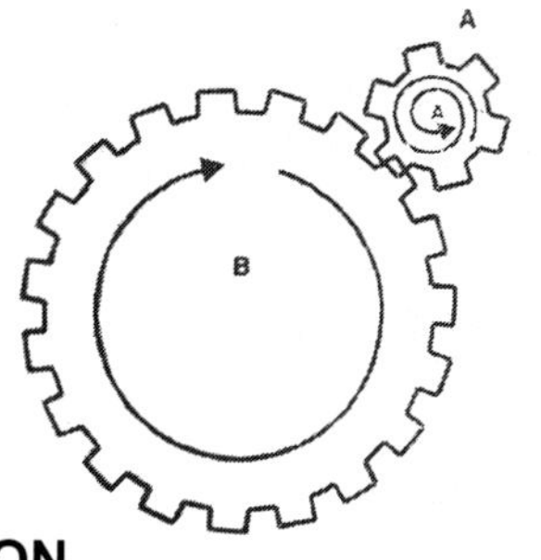

Tractors and Agricultural Machinery

2nd FULLY REVISED AND UPDATED EDITION

Tractors and Agricultural Machinery

2nd FULLY REVISED AND UPDATED EDITION

K. Srinivasan
V.V. Narayanan
Sanjeev Kumar Singh
L. Geethha Lakshmi

NEW INDIA PUBLISHING AGENCY
New Delhi – 110 034

NEW INDIA PUBLISHING AGENCY
101, Vikas Surya Plaza, CU Block, LSC Market
Pitam Pura, New Delhi 110 034, India
Phone: + 91 (11)27 34 17 17 Fax: + 91(11) 27 34 16 16
Email: info@nipabooks.com
Web: www.nipabooks.com

Feedback at feedbacks@nipabooks.com

First edition, 2011
Revised edition, 2016

ISBN: 978-93-85516-10-8

Composed, Designed and Printed in India

Preface Second Edition

A tractor is a vehicle specifically designed to deliver a high tractive effort (or torque) at slow speeds. This is the most important power source used to operate implement, equipment and machinery for numerous agricultural operations. Besides tractors are also put to use for haulage and to pump water from shallow open wells.

The tractor until 1930 resembled a mechanical draft animal, pulling or dragging ploughs, cultivators and equipment. Later, an Irish engineer named Harry Ferguson designed a hydraulic system for 'carrying' implements. Ferguson's system greatly improved traction by transferring the implement's weight onto the back wheels of the tractor.

Over the years many improvements have been made in the design and components to improve the performance and reliability of the tractor. To avoid soil compaction problems, modern tractors are four wheel driven (4WD), with the weight evenly distributed over the four wheels. Modern tractors offer users a plethora of features like global positioning system, load sensing technology, on board computer control units and rollover protection structures.

This second edition contains updated information on working of different sub-assemblies that make a tractor. Uses of tractor for various agricultural and non agricultural operations are vividly described. Besides, updates are also incorporated on various implements, equipment and machinery developed in India for different agricultural operations, *viz.,* land preparation, sowing/planting, weeding, plant protection, harvest, threshing, post harvest and agro-processing. Information on agriculture related sections like special tools and equipment used in horticulture, water lifting devices, calibration of seed drills has also been given.

The first edition was widely used as a standard reference book for graduate students in agricultural engineering and regular engineering colleges. The present edition would also serve the same purpose and can be used as a ready reference for the teaching staff in educational institutions and testing institutions, extension workers, service engineers, scientists and farmers.

Authors

Acknowledgement

The authors wish to express their deep sense of gratitude to Ms. Mallika Srinivasan, Chairperson, Tractors and Farm Equipment Limited (TAFE) for the encouragement and support extended during the preparation of this book. The authors are also indebted to several agricultural universities and institutions coming under the Indian Council of Agricultural Research for providing information on implements, equipment and machinery developed by them for various agricultural operations and also for granting permission to reproduce select photographs.

Thanks are due to many colleagues in the Product Training Centre for their valuable suggestions during the preparation of the first edition of book.

In the present edition, Messers L.G. Raghavan, Antony Vijayaraj and Sathish Kumar from the International Business Unit of TAFE have contributed to new chapters in the Tractors section, to whom we owe our special thanks.

Authors

Contents

Section 1: Tractors

Section 2: Agricultural Machinery

List of Illustrations

List of Tables

Symbols and Abbreviations

Symbols

b	Breadth
cm	Centimetre
CO	Carbon monoxide
^{o}C	Degree Celsius
dia	Diameter
g/kWh	gram per kilowatt-hour
g	Gram
gal	gallon
h	Height
ha	Hectare
HC	Hydro Carbons
hp	horse power
hr	Hour
HV	High Volume
kg	Kilogram
km	Kilometre
kW	kilo Watt
kWh	kilo watt hour
l	Length
lbs	Pound
LH	Left Hand
lit	Litre
LV	Low Volume
m	Metre
min	Minute
mm	Millimetre
N	Newton
NO_x	Oxides of Nitrogen
%	Per cent

P	Particulate number
PM	Particulate Matter
PR	Ply Rating
q	Quintal
Rev	Revolutions
RH	Right Hand
Rs	Rupees
S	Second
t	tonnes and thickness (according to usage)
ULV	Ultra Low Volume
W	Watt
w	Width
WD	Wheel drive
w/m^2	watt per square metre

Abbreviations

AAU	Anand Agricultural University
ANGRAU	Acharya NG Ranga Agricultural University
APAU	Andhra Pradesh Agricultural University
ATRC	Agricultural Tools and Research Centre
BAU	Birsa Agricultural University
Bt	*Bacillus thuringiensis*
CARI	Central Agricultural Research Institute
CCSHAU	Chaudhary Charan Singh Haryana Agricultural University
CIAE	Central Institute of Agricultural Engineering
CIBA	Central Institute of Brackishwater Aquaculture
CIFA	Central Institute of Freshwater Aquaculture
CIFE	Central Institute of Fisheries Education
CIFRI	Central Inland Fisheries Research Institute
CIFT	Central Institute of Fisheries Technology
CIPHET	Central Institute of Post harvest Engineering and Technology
CPCRI	Central Plantation Crops Research Institute
CRRI	Central Rice Research Institute
CTCRI	Central Tuber Crops Research Institute
CTRI	Central Tobacco Research Institute
GAU	Gujarat Agricultural University
GBPUAT	G.B Pant University for Agriculture and Technology
HAU	Haryana Agricultural University
ICAR	Indian Council of Agricultural Research

IICPT	Indian Institute for Crop Processing and Technology
IISR	Indian Institute of Sugarcane Research
KAU	Kerala Agricultural University
MPKV	Mahatma Phule Krishi Vishva Vidyalaya
NBFGR	National Bureau of Fish Genetic Resources
NGO	Non-Governmental Organization
NRC	National Research Centre
NRCM	National Research Centre for Mushroom
NRCOG	National Research Centre for Onion and Garlic
PAU	Punjab Agricultural University
TAFE	Tractors and Farm Equipment Limited
TNAU	Tamil Nadu Agricultural University
UAS	University of Agricultural Sciences
ULV	Ultra Low Volume

Section 1
Tractors

1

Introduction

1.1. Definition

A tractor is a specific vehicle designed to deliver a high tractive torque at slow speeds, for the purposes of haulage or for pulling an agricultural implement or equipment or machinery. This vehicle is also used as a source of power for stand-alone agricultural machineries. The word tractor is taken from Latin being the agent noun of trahere "to pull".

1.2. History of evolution

The first engine-powered farm tractors used steam and were introduced in 1868. These engines resembled small road locomotives. They were used for general road haulage and in particular for timber trade. The most popular steam tractor was the Garrett 4CD.

In 1887, gasoline-fuelled engines were tried as an alternative to steam engines. Later, Henry Ford produced his first experimental gasoline powered tractor in 1907, under the direction of his chief engineer Joseph Galamb. It was referred to as an "automobile plow". While unpopular at first, these gasoline powered machines began to catch on. In the 1910, the size of the tractor was smaller than the earlier introductions and also became more affordable. In 1917, Henry Ford introduced the first mass-produced tractor. These were built in the U.S.A., Ireland, England and Russia. The unique feature of these tractor consisted of the strength of the engine holding the entire machine together. By 1920, tractors with a gasoline powered internal combustions engine had become a norm.

1.3. History in India

The level of mechanization in Agriculture was very low at the time of independence in 1947. But, the initial few five-year plans implemented by the Government of India gave an impetus to the improvement in the production and productivity of major crops including mechanization *via* joint ventures and tie-ups between local industrialists and international tractor manufacturers. It was three decades after independence that the domestic production of 4-wheel tractors grew slowly. In 1990, the tractor production was nearly 2,70,000 units per year and by the 2013, the volume approached to about 6,96,000 units per year. India has presently overtaken the United States as the world's largest producer of four-wheel tractors with over 16 national and 4 multinational corporations engaged in its production. But still the Food and Agriculture Organization estimates that from out of the total agricultural area, less than 50% is under-mechanized land preparation. This indicates that potential for large tractorization still exists. Credit facilities for farmers have been improving steadily over the years. This is one of the prime drivers of the sale of tractor in India.

1.4. Types of tractors

1.4.1. Early designs

Farm tractors introduced in the early twentieth century had a variety of basic front-end types to accommodate different farming needs. These are;

Standard front end: This is known as regular or "wheatland". The models featured a solid front axle that pivoted in the centre and looked like a Horse drawn wagon. Over a period of time, the standard front end became less popular.

Tricycle design: A tricycle was put to use for row crops. This had either a single front wheel, or two front wheels narrowly set together. The front wheels were usually angled towards each other in a 'V' shape pattern so that both wheels fit between rows of crop. The rear wheels had provision for adjustment to accommodate a variety of row spacing. But this design was quite unstable and was prone to rollovers when compared with widely spaced front wheels. The design was also difficult to use with a front-end loader.

High crop tractor: This model featured tall spindles on the front axle and drop-down rear axles. These features provided very high ground clearance. Some manufacturers referred to their high crop tractors as "vegetable" models. High crop tractors had low production volume as compared to the conventional designs. For antique tractor collectors, high crop models often command very high prices.

Over the years, vast changes have taken place in the design of almost all the components of tractors, including size and shape based on the feedback from customers. Many sizes and shapes of tractors are now available for various applications.

1.5. Classification based on usage

1.5.1. Compact Utility Tractor or CUT

This has a small build and is designed chiefly for landscaping and row crops. Typically, this tractor is available in 20 to 40 horsepower range. The compact utility tractor is also equipped with a mid-mounted PTO as well as a standard rear PTO. Like any tractor, this also has a three-point hitch which can be controlled hydraulically. Usually a compact utility tractor is a four wheel drive having hydrostatic transmission and the mandatory rollover protection structure. These tractors are fitted with special smaller implements as compared to the ones required for full size tractors.

1.5.2. Garden tractor

These are small, light and simple tractors designed for use in domestic gardens. These are more sturdily built, with stronger frames, axles and transmissions and capable of mounting other implements such as harrows, cultivators/ rotavators, sweepers, rollers and dozer-blades.

1.5.3. Backhoe loader

This is a variation from the traditional farm tractor. As its name suggests, it has a loader assembly located on the front, as well as a backhoe at the back. These are put to use for construction, light transportation, digging and small demolitions. This is very ideal for urban engineering purposes like construction and repair in places which are too small for the use of large equipment.

1.5.4. Bulldozer

The bulldozer is a strong track-type tractor with blades in the front. Bulldozers have a rope-winch at the back and are specifically used for pushing and dragging things. Many are modified to perform other tasks such as scooping up of rock, earth and similar materials.

1.5.5. Row crop

This type of tractor is designed for planting and cultivating row crops. It is available with wide adjustable front wheels. Modern row crop tractors are available in either two wheel drive or front wheel assist.

1.5.6. Four wheel drive

It has increased traction and pulling power over the traditional two wheel drive tractors. The treads of tyres of the four wheel drive tractors are all of the same size. Both front and rear wheels travel at the same speed. Many four wheel drive tractors have high hp and some have even eight wheels.

1.5.7. Modern tractor

The modern tractor is built to perform several functions. The tractors are fitted with Global Positioning System (GPS) devices and power steering. Many variants in different sizes and horse power have been developed for row and high value crop for heavy field work. Farm equipment and implements are usually attached or hitched to the back end of the tractor through a three point linkage.

1.5.8. Track layer

This is considered a heavy duty tractor which is used for moving soil and heavy equipment on large fields, hillsides and rough land.

1.6. Types of tractor power

A number of terms are widely used to describe power in the tractor. They are Drawbar horsepower (dbhp): It is the power available for the tractor to haul or pull an implement. This is a measured figure rather than a calculated one. To determine the maximum power available, a controllable load is required; it is normally a second tractor with its brakes applied, in addition to a static load.

Brake horsepower (bhp): It is the measure of an engine's horsepower without the loss in power caused by the gearbox, alternator, differential, water pump and other auxiliary components such as power steering pump and muffled exhaust system. The output delivered to the driving wheels is less than that obtainable at the engine's crankshaft.

Power take-off: It is the power generated and measured at the tractor's power take-off shaft. The shaft transfers engine power to the driven implement attached to the tractor.

1.7. Major tractor manufacturers

Table 1.1. Major tractor manufacturers, hp range and brands

Manufactures (Registered or Admin offices)	hp range	Brands
Force Motors Limited, Mumbai-Pune Road, Akurdi, Pune 411 035	27-50	Balwan, Orchard Ox
HMT (International) Limited, HMT Bhavan, 59, Bellary Road, Bangalore 560 032, Karnataka	25-49	HMT
John Deere Equipment Private Limited, Off Pune-Nagar Road, Sanaswadi, Pune 412 208, Maharashtra	35-75	John Deere
Mahindra and Mahindra Limited, 5th floor, Mahindra Towers, G.M. Bhosale Marg, Worli, Mumbai 400 018, Maharashtra	25-85	Mahindra and Mahindra
Mahindra Gujarat Tractor Limited, near Vishwamitri Railway Over bridge, Vishwamitri, Vadodara, 390 011, Gujarat	30-60	Shaktimaan
New Holland Fiat (India) Pvt. Ltd, Tractor Division Plot No. 03, Udyog Kendra, Greater Noida 201 306 Dist. Gautam Budh Nagar, U.P.	32-75	New Holland
PREET Tractors Pvt. Ltd, P.O. Box No. 29, Patiala Road, Nabha 147 201, Punjab	35-70	PREET
Punjab Tractors Ltd, Sas Nagar, Phase-4, Industrial Area, Mohali, Chandigarh 160 055	25-72	Swaraj
Same Deutz-Fahr India (P) Ltd, 72, SIPCOT Industrial Complex (Plot. 72 M), Ranipet 632 403, Tamil Nadu	30-70	SAME
Sonalika International Tractors Limited, Pankaj Plaza-1,Plot No2, Karkardooma, Community Centre, Commercial Complex, Delhi 110 092		
International Tractors Limited, Jalandhar Road, Hoshiarpur 146 022, Punjab	30-90	Sonalika
Standard Combines (P) Ltd, Standard Chowk, Barnala 148 101, Punjab	35-75	Standard Tractor
TAFE Motors and Tractors Limited, Plot 1, Sector D, Industrial Area, Mandideep 462 046, MP	24-60	Eicher
The Escorts Group, Corporate Centre, 15/5 Mathura Road, Faridabad 121 003, Agri Machinery Group, 18/4, Mathura Road, Faridabad 121 007	27-75	Escort, Farmtrac, Powertrac
Tractors and Farm Equipment Limited, Potti Patti Plaza, 77, Nungambakkam High Road, Nungambakkam, Chennai 600 034, TN	25-75	Massey
Ferguson, TAFE, VST Tillers Tractors Limited, P.B.No. 4801, Whitefield Road, Bangalore 560 048, Karnataka	<20	VST and Mitsubishi-Shakti

1.8. Production and sale of tractors

The year wise production and sale of tractors are presented (Table 2). Around 0.3 million tractors are sold every year in recent times. The demand is also steadily growing. The sale of tractors declined only in 2002 and 2003 due to drought. In 2012-13, there was a marginal decline.

Table 1.2. Year-wise production and sale of tractors (in numbers)

Year	Production	Sale
2000-01	NA	249245
2001-02	NA	215559
2002-03	NA	171622
2003-04	191633	191673
2004-05	249077	247531
2005-06	296994	292880
2006-07	352368	352781
2007-08	345762	346501
2008-09	339510	342836
2009-10	433207	440230
2010-11	548397	545109
2011-12	639896	607251
2012-13	578690	590672
2013-14	696801	696828

Source: Tractors Manufacturers Association

1.9. Total number and percentage share of different HP tractors

Different sizes of tractors versus sale are also provided (Table 3). The data reveals that among the tractors manufactured in India ranging from < 20 hp to >50 hp, the latest trend (2013-14) appeared to be between 41 and 50 hp range.

1.10. Total number and percentage share of tractors in different states

The sale of tractors is the highest in Uttar Pradesh followed by Madhya Pradesh, Rajasthan and Maharashtra for the year 2013-14 (Table 4).

Table 1.3. Total number and per cent share of different hp tractors

hp range	Total number and per cent share of different hp tractors					
	2008-09	2009-10	2010-11	2011-12	2012-13	2013-14
<20	2329	3761	4729	7033	18468	22314
	(0.69%)	(0.85%)	(0387%)	(1.16%)	(3.13%)	(3.25%)
21-30	49986	66024	70194	75748	37038	48542
	(14.58%)	(14.99%)	(12.88%)	(12.47%)	(6.27%)	(6.97%)
31-40	157602	202045	226545	250701	238845	224452
	(45.97%)	(45.88%)	(41.56%)	(41.26%)	(40.44%)	(32.21%)
41-50	84927	105652	154191	168695	214766	319444
	(24.77%)	(23.99%)	(28.29%)	(27.76%)	(36.36%)	(45.84%)
>50	47992	62849	89450	105481	81555	72560
	(13.99%)	(14.27%)	(16.41%)	(17.36%)	(13.81%)	(10.41%)
Total	342836	440331	545109	607658	590672	696828

Source: Tractors Manufacturers Association

Table 1.4. Total number and per cent share of tractors in different states

States	Total number and per cent share of tractors in different states					
	2008-09	2009-10	2010-11	2011-12	2012-13	2013-14
AP	38365 (11.25%)	34366 (7.81%)	44099 (8.09%)	41958 (6.90%)	30765 (5.21%)	44163 (6.34%)
Assam	1588 (0.47%)	3583 (0.81%)	3893 (0.71%)	5377 (0.88%)	4600 (0.78%)	4969 (0.71%)
Bihar	17511 (5.13%)	29050 (6.60%)	27244 (5.00%)	29162 (4.80%)	34861 (5.90%)	40437 (5.80%)
Jharkhand	3643 (1.07%)	5669 (1.29%)	6524 (1.20%)	6697 (1.10%)	6756 (1.14%)	8349 (1.20%)
Gujarat	20164 (5.91%)	24291 (5.52%)	40951 (7.51%)	56664 (9.32%)	43245 (7.32%)	46669 (6.70%)
Haryana	22532 (6.60%)	28645 (6.51%)	24122 (4.43%)	26040 (4.29%)	27900 (4.72%)	35290 (5.06%)
HP	907 (0.27%)	1201 (0.27%)	1404 (0.26%)	1398 (0.23%)	11255 (0.21%)	1530 (0.22%)
J and K	1115 (0.33%)	1603 (0.36%)	2128 (0.39%)	2618 (0.43%)	3577 (0.61%)	3792 (0.54%)
Karnataka	13746 (4.03%)	23904 (5.43%)	28403 (5.21%)	31533 (5.19%)	23741 (4.02%)	30599 (4.39%)
Kerala	374 (0.11%)	577 (0.13%)	671 (0.12%)	631 (0.10%)	570 (0.10%)	385 (0.06%)
Maharashtra	25381 (7.44%)	34184 (7.77%)	53604 (9.83%)	57191 (9.41%)	44221 97.49%)	65405 (9.39%)
Madhya Pradesh	24149 (7.08%)	33345 (7.57%)	48435 (8.89%)	50597 (8.32%)	70822 (11.99%)	87831 (12.60%)
Chattisgarh	8725 (2.56%)	10167 (2.31%)	12861 (2.36%)	13277 (2.18%)	15831 (2.68%)	24277 (3.48%)
Orissa	5099 (1.49%)	7902 (1.79%)	11134 (2.04%)	9254 (1.52%)	9221 (1.56%)	12139 (1.74%)
Punjab	19986 (5.86%)	29010 (6.59%)	26621 (4.88%)	25987 (4.28%)	28260 (4.78%)	30187 (4.33%)
Rajasthan	25743 (%)	31808 (7.23%)	35173 (6.45%)	53175 (8.75%)	63275 (10.71%)	74920 (10.75%)
Tamil Nadu	14334 (4.20%)	15810 (3.29%)	20638 (3.79%)	26298 (4.33%)	18448 (3.12%)	9865 (1.42%)
Uttar Pradesh	51312 (15.04%)	73670 (16.73%)	76981 (14.12%)	82613 (13.60%)	84559 (14.32%)	95650 (13.73%)
Uttarkhand	186 (0.05%)	2321 (0.53%)	2430 (0.45%)	2057 (0.34%)	2099 (0.36%)	2408 (0.35%)
West Bengal	6096 (1.79%)	9188 (2.09%)	13045 (2.39%)	12187 (2.01%)	12752 (2.16%)	13756 (1.97%)
Others	1983 (0.58%)	2310 (0.53%)	1876 (0.34%)	2172 (0.36%)	876 (0.15%)	1220 (0.18%)
Export	38214 (11.20%)	37622 (8.55%)	62872 (11.53%)	70772 (11.65%)	62890 (10.65%)	62677 (8.99%)
Total	341153	440230	545109	607658	590672	696828

Source: Tractor Manufacturers Association

1.11. Chapters in this section

All the sub-assemblies that constitute a tractor are covered in detail. The examples cited in the text relate to those found in the common tractors manufactured in India. The information provided in "Trouble shooting" for each of the sub-assemblies also refers to all makes of tractors and models.

2

Engine and Associated Systems

2.1. Introduction

The engine is the source that generates power. It transforms chemical energy in fuel into mechanical energy for power. Diesel internal combustion engines are invariably used in all the tractors. In these engines, very high compression of the air ignites the fuel as compared to spark plug that ignites fuel in petrol engines. The air in the diesel engine is compressed with a compression ratio typically between 15 and 22 resulting in a 40 bar (about 600 psi) pressure as compared to 8 to 14 bar (about 200 psi) in the petrol engine. This high compression heats the air to 550°C (1022°F). At about this moment, small droplets of fuel are sprayed in an atomized form directly into the compressed air in the combustion chamber.

2.2. How does it work?

Any engine will work with four strokes, namely suction, compression, power and exhaust (Fig 2.1). It will also complete two revolutions of the crankshaft. In the suction stroke, pure air is drawn into the cylinder through the inlet valve. The exhaust valve gets closed and the crankshaft completes half a revolution. In the compression stroke, the piston moves up while the valves remain closed. The air which is drawn into the cylinder during the suction stroke is compressed. There is a rise in temperature. The hot air ignites the fuel injected into the cylinder. The crankshaft again completes half a revolution. In the power stroke, the hot products of combustion, *viz.,* carbon dioxide and nitrogen expand. The piston is forced down. The valves remain closed and the crankshaft completes half a revolution. In the exhaust stroke, the exhaust valve opens while the piston reaches the lowest point of travel. Then the burnt gases are swept out.

The crankshaft completes the final half revolution. The energy produced during the power stroke is stored by the fly wheel and used for the other three strokes and power output.

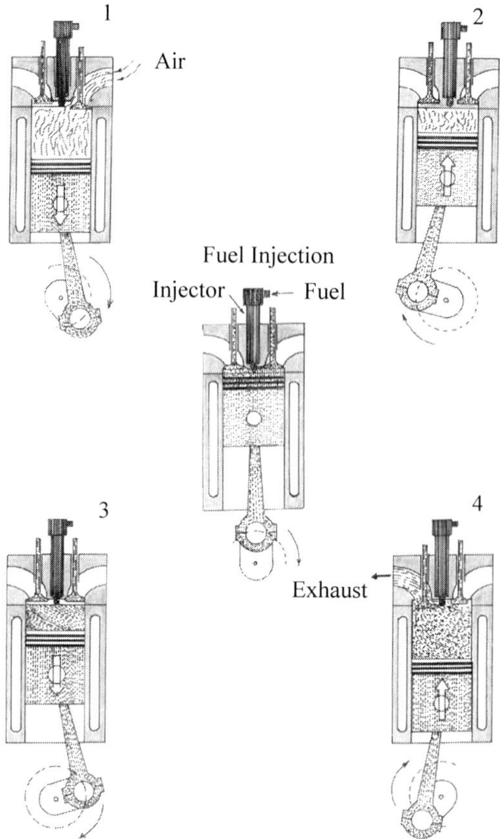

Fig 2.1. Engine four strokes - A complete cycle

2.3. Systems in the engine

2.3.1. Fuel system

The fuel system consists of the following, viz., fuel tank, water separator, feed pump, hand primer, primary and secondary filters and FIP (Fuel Injection Pump) with governors.

Fuel tank

The capacity of the fuel tank in tractors varies between 35 and 70 litres depending on the size and capacity of the tractor. The maintenance of the fuel tank involves draining of fuel from the tank and cleaning after every 750 hrs of run. Generally the fuel cock is located below the diesel tank. The gauze filter in the cock also requires cleaning.

Water separator

At times, the diesel gets contaminated with water. Since water is heavier than diesel, it gets separated and collects at the bottom of the separator. Maintenance involves loosening the drain screw at the bottom of the bowl to drain the water on sighting the red float reaching the set mark on the separator. The frequency may be increased if the contamination of water in the fuel is excessive.

Feed pump

The feed pump is a reciprocating pump and gets drive from the camshaft that gives drive to the fuel injection pumps. The feed pump draws fuel from the fuel tank through a sedimentary bowl and micro filter and pumps it under low pressure, through the fuel filters into the suction gallery of the injection pump. Its maintenance involves cleaning of the preliminary (micro) filter and the sediments. The plunger surface of the pump might dry if the tractor is not used for a long time. In such cases, filling some fuel or oil through the suction valve (on the inlet end) prior to use facilitates suctioning of the feed pump. Periodical examination of the rubber sealing washer in the sedimentary bowl is also necessary. This may require replacement if it is distorted or has become too hard.

Hand primer

The hand primer is usually screwed to the feed pump above the suction valve. The fuel is pumped from the fuel tank through the water separator to the fuel injection pump when the engine is at rest. To operate the priming device, the knurled knob is screwed out until the plunger is pulled upwards. This causes the suction valve to open and enables fuel to flow into the suction chamber. When the plunger is pressed down, the suction valve closes while the pressure valve opens. This enables the fuel flow through the feed pipe and filter into the injection pump. After use, it is essential to screw the knob completely before starting. The hand primer is provided to bleed the air trapped in the closed fuel circuit. This air gets trapped into the system when the fuel in the system is fully used/ drained. Bleeding the air is carried out by loosening the purging screws on top of the primary and secondary filters and operating (pumping) the hand primer.

Filters

Filters effectively remove foreign particles, viz., dusts, scale, rust and water (to some extent) from the fuel. Solid impurities have an obstructive or plugging effect on delivery valves and tiny spray holes of the nozzle of the fuel injector.

The primary filter consists of a cloth like material coiled around a perforated tube. The top and bottom are sealed by end covers. Fuel flows radially from the outside to the inside. The secondary filter is either a coil type or star type. In the coil type, the filter paper is wound around a tube and neighbouring layers are glued at the top and bottom by end covers. The fuel flows from the bottom to the top. The dirt particles are held back in filter pockets. In star-type filter, the fuel flows radially from outside to inside. The filtered fuel collects inside the perforated centre tube and flows upwards. The dirt particles are trapped on the filtering surface or settle down at the bottom. The primary filter inserts should be changed every 500 hr. The secondary filter inserts should be changed every 750 hr. The filter inserts may get clogged due to asphalt or waxy compounds present in the fuel. In such cases, these have to be changed earlier than the specified hours. The filter inserts should never be cleaned and reused.

Governors

The governor regulates the quantity of fuel injected in to the engine. Mechanical and pneumatic type governors are fitted in diesel engines. The mechanical type is mounted on the fuel injection pump. The control rack of the fuel injection pump is connected to the governor through a flexible joint and the governor controller is connected to the accelerator pedal. There are two types of pneumatic governors, *viz.*, venturi control and diaphragm unit.

The venturi unit is mounted on the engine induction manifold at the inlet side. Air, which is drawn into the engine through an air filter attached to this unit, passes to the venturi. A butterfly valve and a connection to the vacuum line are located at the narrowest point of the venturi. The butterfly valve is connected to the accelerator pedal through a control lever and the linkage. Two setting screws are provided to set the idling and maximum speed in the tank manifold. In the diaphragm unit, it is mounted on to one of the end faces of the fuel injection pump. This unit is subdivided by a leather diaphragm into an atmospheric chamber and a vacuum chamber. The atmospheric chamber is always maintained under atmospheric pressure through an air filter. The diaphragm is under pressure from the governing spring control rod. Difference in the pressure on either side of the diaphragm controls the movement of the diaphragm. The pneumatic type governors have almost become obsolete and only mechanical governors are used in many products.

2.3.2. Air induction system

In this system only fresh air is allowed into the engine. The various types of filters used are dry type, wet type, dry and wet type and four stage. In a four stage filter (Fig 2.2), the air that enters the air pre-cleaner is cleared of solid and heavy dust particles (pre-cleaning). The smaller dust particles get separated when they hit the oil in the oil bath before passing on to the layers of wire mesh, paper filter and felt filter. Maintenance involves cleaning of pre-cleaner and oil bath every 10 hours of operation. The frequency of cleaning needs to be increased when working in a dusty atmosphere, *viz*, brick kiln, coal mines, *etc*. The other filters are cleaned during regular service periods as recommended by the Original Equipment Manufacturer (OEM).

Fig 2.2. Four stage air filter - Assembled and individual components

2.3.3. Exhaust system

The exhaust system expels the burnt gases. The main components consist of exhaust manifold and silencer. It is advisable to keep the exhaust outlet pipe closed to prevent water entry when the engine is put to rest for longer periods.

2.3.4. Cooling system

All diesel engines require a temperature of 70 to 95° C to perform at their optimum efficiency. The reasons are;

(a) The atomized fuel gets fully burnt.

(b) All the components in the cooling system get expanded to the required size so that they can run together without too much wear and tear.

(c) The lubricating oil attains proper viscosity and reaches all the components.

The water flowing through the engine cools and controls the engine temperature. The system comprises four major components - radiator, thermostat, fan and water pump. The belt driven pump circulates the water. The thermostat compels the water to bypass the radiator, thus allowing the water to reach the normal working temperature. Once the working temperature is reached, the thermostat allows the water to flow through the radiator where it is cooled by the passage of air through the radiator. Cooling is further assisted by the fan. The entire system requires examination when the coolant boils or the engine becomes too cool.

2.3.5. Lubrication system

Whenever two metal surfaces-such as a shaft in a bearing or a piston in a cylinder wall, rub against each other, friction is caused and heat is generated. The mating surface might break though they are highly polished. To prevent these, a thin film of oil must be maintained to minimize the amount of friction, wear and heat generated either between rotating or sliding parts of the engine under all operating conditions. Hence, an adequate amount of oil must be fed continuously to those parts of the engine that requires lubrication. The following parts of the engine have to be lubricated, namely crankshaft bearing, crank pin, big and small ends of the connecting rod, bushes of the gudgeon pin, inner wall of the cylinder, piston rings, valve operating mechanism and timing gears. The oil passing through the liners also requires cleaning since carbon is formed during burning of fuel.

Oil level check

The correct level of clean engine oil must be maintained in the oil sump. It is ensured that the tractor is on a level ground and sufficient time has elapsed to allow oil to drain back to the sump. The dipstick is withdrawn, wiped clean and inserted into the sump. The dipstick is again removed and held vertically to check whether oil is present between the high and low level marks. Oil is to be replenished only when the level goes below the low level mark. It is desirable to maintain the level midway between the maximum and minimum level.

The oil level gets settled down to a point after 150 to 200 hours of initial change. Subsequently, an amount of about 0.5% of fuel consumption is allowed. Drastic consumption of lubricating oil requires detailed examination of the entire engine to identify the exact problem.

Changing the oil

The tractor must be parked on level ground and an oil container placed beneath the drain plug of the sump. The drain plug and the seal are removed. Then the old oil is allowed to drain until the sump is completely empty. Afterwards, the drain plug is put in position along with a new seal. The sump is filled with new oil of the correct grade midway between the high and low level marks on the dipstick.

The lubricating oil canister from the filter head is also removed and filter element discarded. The filter housing is cleaned and filled up with the correct grade of new oil. During the process, the filter seals are lubricated and sufficient time is allowed for the oil to soak through the filter elements. Afterwards, the newly oiled filter head is screwed until the seal just touches the head. After renewing the oil and filter, the engine is allowed to idle and examined for leaks. When the engine stops, sufficient time is allowed for the oil to drain to the sump. The oil level is re-checked again with a dipstick. Oil is then replenished in the sump to bring the level up to the full level mark on the dipstick.

2.4. Major components and their functions (Figs 2.3. and 2.4.)

Components	Functions
Cylinder	Main support for all the engine parts. Crankcase and cylinder blocks are machined together to achieve a perfect match.
Cylinder liner	Integral part of cylinder block. Fits into cylinder block. Withstands high working temperature and pressure. Resistant to corrosion and wear.
Cylinder head	Forms top of the cylinder/engine. Holds valves, valve guides, tappets, valve springs, manifold and front end. Supports rocker assembly. Provides passage for the cooling water and lubricating oil. Serves as a passage for incoming air and exhaust gas
Piston	Helps transmission of explosion force to the crankshaft through the connecting rod. Has maximum load carrying capacity with minimum weight. Possess maximum heat withstanding capacity with minimum expansion
Piston pin	Also known as gudgeon pin. This connects the piston to the connecting rod and provides a bearing for the connecting rod to pivot upon as the piston moves
Piston assembly	Transmits force to crankshaft through the connecting rod. Serves as a guide for the upper end of the connecting rod
Compression rings	Provides sealing wall pressure. Conducts heat as rapidly as possible. Prevents leakage of gases from the combustion chamber to the sump.

Oil scrapper rings	Scrapes excessive oil and carbon formations
Fuel injection pump	Pumps fuel as per demand by accelerator/governor in correct quantity and at the right time through injectors
Injectors	Atomises the fuel for easy burning
Rocker Assembly	Assists in the opening and closing of valves. Provides passage for lubricating oil.
Water pump	Assists in circulation of water for cooling the engine
Timing gear	Transmits the drive from the crankshaft to camshaft, lubricating oil pump and fuel injection pump
Flywheel	Stores energy and maintains uniform engine speed by carrying the crankshaft through the intervals when it is not receiving power from a piston. Supports the clutch assembly
Sump	Acts as reservoir for lubricating oil
Lubricating oil pump assembly	Supplies lubricating oil under pressure to all moving parts of the engine, *viz.* crankshaft, connecting rod, bearings, camshaft and rocker assembly
Crank shaft	Converts linear motion of the piston into rotational motion of the fly wheel. The crankshaft is further connected to the flywheel, clutch, main drive shaft of the transmission, torque converter and belt pulley.

Fig 2.3. S series Simpson Engine - Cross section

1-Web crankshaft 2-Oil gallery 3-Auxiliary drive 4-Timing case 5-Connecting rod 6- Piston 7- Cylinder block 8- Ring groove 9- Exhaust valve 10- Cylinder head 11- Valve guide 12- Valve spring 13- Rocker arm 14- Inlet valve

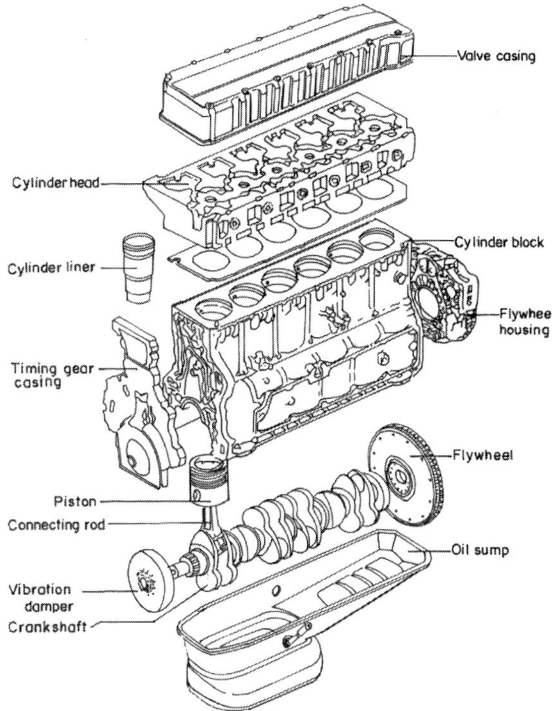

Fig 2.4. Engine - Major components

2.5. Number of cylinders versus power

The tractor engines are built with either one or two or three or four cylinders. The size of the tractors, *i.e.* horse power has no relationship with number of cylinders in them. Almost all the engines are four stroke models.

2.6. Direct injection engines

All the modern tractors are fitted with Direct Injection (DI) engines. The main aspect that defines a DI engine is the application of the fuel directly into the combustion chamber. The older versions of diesel engines used port fuel injectors, where fuel was applied in the intake ports upstream of the intake valve. The port fuel injection is less precise since it just sprays fuel into the intake port, which then mixes with the air in the port and rushes into the combustion chamber when the intake valve opens. The DI fuel application is a big leap forward. It determines the precise timing of the fuel entering the combustion chamber and opens up a host of opportunities for engine tuners to make power, reduce emissions and increase the durability of the engines, all at the same time. Tractors fitted with DI engines are usually run at a minimum rpm of 700-750.

2.7. Diesel engine versus gasoline engine

Diesel engines have several advantages over other internal combustion engines. They produce more torque than one can get from petrol engines. Also, diesel engines have a much longer life than petrol engines if properly maintained. They burn less fuel than a petrol engine performing the same work, due to the engine's high efficiency and can deliver much more of their rated power on a continuous basis than a petrol engine.

Diesel fuel is considered safer than petrol in many applications. This fuel will not explode and does not release a large amount of flammable vapour. Diesel fuel is cheaper as compared to petrol.

2.8. Supercharged and turbocharged engines

Most diesel engines are now turbocharged and some are both turbocharged and supercharged (Fig 2.5.) depicts advantages and disadvantages of each of the engines mentioned in the following paragraphs). Because diesel engines do not have fuel in the cylinder before combustion is initiated, more than one bar of air can be loaded in the cylinder without pre-ignition. A turbocharged engine can produce significantly more power than a naturally aspirated engine of the same configuration. Storage of more air in the cylinders allows more fuel to be burnt resulting in the production of more power. A supercharger is powered mechanically by the engine's crankshaft, while a turbocharger is powered by the engine exhaust, not requiring any mechanical power. Turbo charging improves the fuel economy of diesel engines by recovering waste heat from the exhaust. The engine output is also increased to a significant extent. Because turbocharged or supercharged engines produce more power for a given engine size as compared to naturally aspirated engines, attention must be paid to the mechanical design of the components, lubrication and cooling to handle the power. Pistons are usually cooled with lubrication oil sprayed at the bottom of the piston.

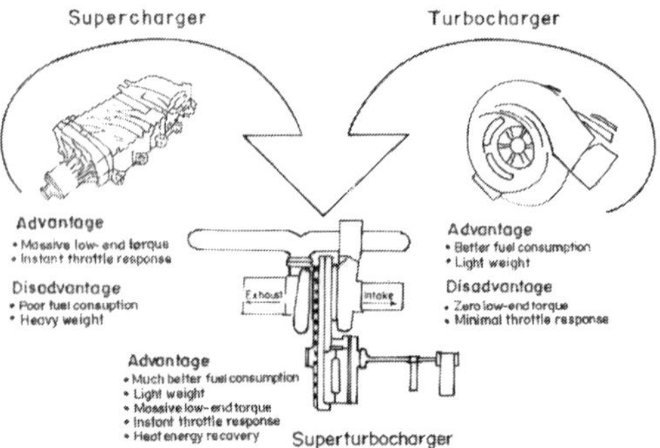

Fig 2.5. Engine - Supercharged and turbocharged

2.8.1. Working principle of a turbocharger

To achieve boost to the engine, the turbocharger uses the exhaust flow from the engine to spin a turbine, which in turn spins an air pump. The turbine in the turbocharger spins at speeds of up to 1,50,000 rotations per minute (rpm): that is about 30 times faster than most car engines. Since it is hooked to the exhaust, the temperatures in the turbine are also very high.

2.8.2. Working principle of a supercharger

A supercharger is an air compressor that increases the pressure or density of air supplied to an internal combustion engine. This gives each intake cycle of the engine more oxygen, letting it burn more fuel and do more work, thus increasing power.

Power for the supercharger can be provided mechanically by means of a belt, gear, shaft or chain connected to the engine's crankshaft. Common usage restricts the term supercharger to mechanically driven units.

2.8.3. Types of turbocharger

Twin turbo

Twin-turbo designs have two separate turbochargers operating either in a sequence or in parallel. In a parallel configuration, both turbochargers are fed one-half of the engine's exhaust.

Sequential Turbo

In a sequential turbo one turbocharger runs at low speeds and the other turns on at a pre-determined engine speed or load. Sequential turbochargers further reduce turbo lag, but require an intricate set of pipes to properly feed both turbochargers.

Two stage variable twin-turbo

Two stage variable twin-turbo employs a small turbocharger at low speeds and a large one at higher speeds. They are connected in a series so that pressure boosted from one turbocharger is multiplied by another, hence the name "two stage" The distribution of exhaust gas is continuously variable. Hence, the transition from using the small turbocharger to the large one can be done incrementally.

2.9. Fault diagnosis, causes and solutions

Problems in the engine can be caused due to many reasons. It is advisable for a tractor owner to visit an authorized dealership to rectify the problems. Diagnosis of the problem involves years of experience and use of common sense. These are the common troubles faced by tractor owners of any make pertaining to the engine.

2.9.1. Engine

Fault	One or few of the under mentioned causes
Erratic running	Faulty stop control operation. Blocked fuel feed pipe. Faulty fuel feed pump. Clogged filter inserts. Restriction in air cleaner. Faulty feed injection pump. Leaky high pressure pipe joints. Incorrect tappet adjustment. Incorrect pump injection timing. Incorrect pump phasing/unequal delivery. Worn out pump control rack/toothed quadrant. Engine compression poor/partially seized. Piston seizure. Overheating
Starting trouble	Weak battery/defective starter. Glow plugs/Defective circuit. Low cranking speed. No fuel in the tank. Faulty stop control operation. Sticky/leaky overflow valve. Air in the fuel system. Governor stop lever not fully disengaged. Worn out injection pump/feed pump. Poor/ partially seized engine. Sticky valves/Broken springs /clearance
Knocking sound	Incorrect pump injection timing. Incorrect pump phasing /Unequal delivery. Delivery valve in pump is stuck/ broken spring. High opening pressure. Nozzle stuck in the open position. Faulty fuel feed pump. Faulty atomizers or incorrect type. Faulty cold starting equipment (Thermostat). Overheating. Incorrect tappet adjustment. Worn cylinder bores. Piston seizure. Incorrect piston height. Clogged spray holes, sticky needle, poor spray. Poor/partially seized engine compression
Lack of power	Blocked breather hole in the tank. Blocked low pressure pipes. Clogged filter inserts. Sticky/leaky overflow valves. Air in the fuel system. Loose/restricted accelerator linkage. Incorrect timing of injection pump. Maximum speed setting too low. Worn out injection pump/feed pump. Incorrect injection timer advance. Leaky high pressure pipe joints. Restricted high pressure pipes. Incorrect position of nozzle. Poor or partially seized engine compression. Overloaded engine /vehicle. Excessive blow-by
Excess fuel consumption	Restriction in air cleaner or induction system. Air in fuel system. Faulty fuel injection pump. Incorrect pump injection timing. Defective excess fuel device. Idling speed setting too high. Full load/ course of delivery too high. Incorrect injection timer advance. Clogged spray holes. Excessive nozzle needle lift. Clogged air filter/ exhaust system. Partially seized engine compression. Valves/Broken

	springs/Incorrect clearance. Overloaded engine/vehicle. Too long engine idling. Overheating. Cold running. Worn cylinder bores
Black smoke	Incorrect tappet adjustment. Incorrect type or grade of fuel. Incorrect pump injection timing. Incorrect valve timing. Incorrect use of cold start equipment. Pump delivery valve is stuck/springs broken. Faulty fuel pump. Faulty atomizers. Incorrect injection timer advance. Incorrect position of nozzle. Low opening pressure. Nozzle stuck in open position/Broken NH spring. Sticky valves. Broken, worn or sticky piston rings. Clogged spray holes, Sticky needle. Clogged air filter/exhaust system. Poor/partially seized engine compression
Overheating	Restriction in exhaust pipe. Leaky cylinder head gasket. Full load/ course of delivery too high. Spray holes clogged. Faulty fuel pump. Faulty atomizers or incorrect type. Poor/partially seized engine compression. Excessive carbon deposits in combustion. Faulty cooling/lubricating oil system. Piston seizure. Overloaded engine/ vehicle

2.9.2. Piston

Fault	One or few of the under mentioned causes
Uneven bearing on the skirt	Misaligned connecting rod and piston bearing unevenly in the bore of the cylinder
Abrasive wear on the skirt	Fine and hard dust particles entering the bore with air getting trapped between the piston and cylinder, cutting fine grooves in the form of scratches on the skirt surface during normal working. This is accompanied by a similar wear on the rings and cylinder wall
Failure of the circlip groove	Excessive end float of the gudgeon pin causing it to hit the circlip and damage the groove end. Careless installation of circlip causes damage to the groove
Uneven bearing on the gudgeon pin	Improper gudgeon pin fit
Breakage of top crown	Leakage of water on top of the piston builds up into a small reservoir. When it exceeds, the clearance volume a hydraulic lock is created during the compression strokes. This breaks the crown of the piston

2.9.3. Fuel pump

Fault	One or few of the under mentioned causes
Pump does not deliver fuel	Empty fuel tank. Closed fuel tank. Closed fuel inlet pipe. Dirty filter element. Air lock in pump. Jammed or leaking delivery valve
Pump does not deliver fuel uniformly	Air lock in pump. Broken delivery valve spring. Damaged delivery valve seat/face. Pump plunger remains suspended in the barrel. Dirty inlet pipe or filter element. Head between the tank and pump too small

Jammed control rod	The pump plunger as seized or the control rod toothed rack is coated with dirt
Quantity of fuel delivered at every stroke is insufficient	Leaky delivery valve. Leaky joints in the pressure system
Excessive fuel consumption	Incorrect setting of governor. Leakage of fuel. Wrong pump timing. Excessive load on engine. Incorrect valve timing. Poor compression

2.9.4. Nozzle failure

Fault	One or few of the under mentioned causes
Injection pressure too high	Incorrect setting of injection pressure. Injection nozzle valve sticks or is seized in the nozzle body
Injection pressure too low	Incorrect setting of injection pressure. Broken nozzle holder spring or sticking nozzle needle
Nozzle dribbles. Spray from nozzle is deformed	Incorrect seating of nozzle valve

2.9.5. Lubrication system

Fault	One or few of the under mentioned causes
Oil leaks	Level of lubricant too high. Damaged or improperly fitted gaskets. Dummy plug of planetary rail not sealed
Oil leak through cup gear shift lever	Defective/choked breather fitted on hydraulics lift cover
Noise from gear box in neutral	Worn out bearing. No lubricant. Worn out gears with broken or chipped teeth. Too much backlash in the gear train.
Noise from the gear box when the tractor is in motion	Noises within the clutch or in the rear end may also seem to come from transmission
Engine working and gear in neutral position.	Humming noise from gear box when clutch pedal is depressed Defective clutch release bearing
Intermittent noise from the gear box with engine running and gears in neutral	High spots in constant mesh of gears and input shaftCheck whether this is hot spot
Noise while changing gears	Excessive play in the clutch pedal
Gear slips	Weak or broken plunger spring. Insufficient meshing of gear

2.9.6. Cooling system

Fault	One or few of the under mentioned causes
Coolant boils	Insufficient water in the radiator. Leaking radiator cap. Leaking hoses and joints. Weak or broken spring. Defective valve seat in radiator cap. Fan blades incorrectly fitted. Slack or worn fan belt. Faulty thermostat (remains closed or not opening properly). Perished cooling system hoses. Choked radiator core or restricted passages. Damaged or corroded water pump impeller. Radiator choked with mud and chaff
Engine runs too cool	Faulty thermostat (remains closed or not opening properly)

2.9.7. Air induction and exhaust system

Fault	One or few of the under mentioned causes
Clogged exhaust pipe leading to overheating of engine and associated engine problems	Excessive carbon deposits
Engine problems	Excessive accumulation of dirt

2.10. Blow-by

Blow-by occurs when the explosion in the engine's combustion chamber causes fuel, air and moisture to be forced past the rings into the crankcase. The rings should be maintained in an excellent condition in order to prevent blow-by.

2.10.1. Effects of blow-by

Blow-by inhibits performance because it results in a loss of compression. When the expanding gases slip past the rings they cannot as effectively push the piston down and make the vehicle go. As a result, the tractor will have less horsepower. This also results in the loss of fuel economy. When the fuel, air and moisture slip into the crankcase they contaminate and dilute the oil in the crankcase

2.11. Measurement of engine

Engine is measured in different ways, *viz*., horsepower, torque, thermal efficiency, mechanical efficiency, volumetric efficiency and displacement.

Horse power: It is rate at which at which work is done. Needless to mention, work is a product of distance and weight. One horsepower is equivalent to 33,000 ft-lb/ minute. Gross horsepower is the maximum horsepower developed by engine equipped with accessories needed for operation. Many a times, brake horsepower is used to denote the measure of actual usable horsepower delivered at crankshaft.

Torque: It is the twisting force and is measured by the product of force and distance in pound feet. Engine torque is the actual measurement of engine's true power. Peak torque in an engine is reached at engine's maximum volumetric efficiency.

Thermal efficiency: It is the measurement of energy provided by burning fuel to produce horsepower. Heat is usually lost in cooling, lubrication and exhaust systems. Thermal efficiency of an average engine is about 25%.

Mechanical efficiency: It is the relationship of power developed within engine and power delivered at crankshaft. Most engines have mechanical efficiencies of about 90%.

Volumetric efficiency: It is the measurement of engine's ability to draw air and fuel into cylinders and is determined by the ratio of actual amount drawn in and actual cylinder volume. As the speed of engine increases, there is decrease in volumetric efficiency. This efficiency is hence affected by speed of engine, intake design, temperature, valve size and exhaust design.

Displacement: It is the total number of cubic inches or liters of space in a cylinder when the piston moves from the top dead centre to the bottom dead centre. It is calculated as bore x stroke x number of cylinders. More displacement results in more power.

For easy understanding by a common man, engine measurement is denoted in cc (cubic centimetre). It is the total volume displaced by the cylinders of the engine through one revolution.

2.12. Emission standards

Emission standards are requirements that set specific limits to the amount of pollutants that can be released into the environment. Many emission standards focus on regulating pollutants released from different types of vehicles.

Bharat (Trem) Emission Standards for tractors						
Engine Power	Date	CO	HC	HC+NO_x	NO_x	PM
kW		g/kWh				
Bharat (Trem) Stage I						
All	1999.10	14.0	3.5	-	18.0	-
Bharat (Trem) Stage II						
All	2003.06	9.0	-	15.0	-	1.00
Bharat (Trem) Stage III						
All	2005.10	5.5	-	9.5	-	0.80
Bharat (Trem) Stage III A						
$P < 8$	2010.04	5.5	-	8.5	-	0.80
$8 \leq P < 19$	2010.04	5.5	-	8.5	-	0.80
$19 \leq P < 37$	2010.04	5.5	-	7.5	-	0.60
$37 \leq P < 75$	2011.04	5.0	-	4.7	-	0.40
$75 \leq P < 130$	2011.04	5.0	-	4.0	-	0.30
$130 \leq P < 560$	2011.04	3.5	-	4.0	-	0.20

Acronyms: CO: Carbon monoxide, g/kWh: gram per kilowatt-hour, HC: Hydro Carbons, NO_x: Oxides of Nitrogen, P: Particulate number, PM: Particulate Matter

3.1

Clutch

3.1.1. Introduction

Clutch is a device which is used to connect and disconnect the tractor engine from the transmission gears and drive wheels. It is fitted in between the engine and gear box. The clutch transmits power by means of friction between driving members and the driven members.

3.1.2. Necessity of clutch in a tractor

A clutch in a tractor is essential for the following reasons:

(i) The engine needs cranking by any suitable device. For easy cranking, the engine is disconnected from the rest of the transmission unit by a suitable clutch. After starting the engine, the clutch is engaged to transmit power from the engine to the gear box.

(ii) To change the gears, the gear box must be kept free from engine power; otherwise, the gear teeth will be damaged and the engagement of the gear will not be perfect. This work is done by a clutch.

(iii) When the belt pulley of the tractor works in the field, it needs to be stopped without stopping the engine. This is done by a clutch.

3.1.3. Essential features of a good clutch

(i) It should have good ability to take load without dragging and chattering.

(ii) It should have a higher capacity to transmit maximum power without slipping.

(iii) The friction surface should be highly resistant to heat effect.

(iv) The control by hand lever or pedal lever should be easy.

3.1.4. Types of clutches

Friction clutch

Friction clutch is the most popular in four-wheel tractors. A friction clutch produces gripping action, by utilizing the frictional force between two surfaces. These surfaces are pressed together for transmission of power. While starting the engine, the clutch pedal is depressed. After the start of the engine, the clutch pedal is slowly released to increase the pressure gradually on the frictional surfaces until there is no slip. Thus, the driven plate is gripped firmly to the driving plate. The transmission of power depends upon the kind of material used for the friction members and the intensity of the force pressing them together.

Friction clutches may be subdivided into three classes: (i) single plate clutch or single disc clutch (ii) multiplate clutch or multiple disc clutch and (iii) diaphragm clutch

(i) Single plate clutch

Function: It transmits power from the engine to the gear box as well to the hydraulics together or cuts it off; i.e. power from the engine is either provided to both units (transmission and hydraulics) or cut off. This may be called a single disc clutch. There is only one clutch plate in this type. The clutch plate is pressed against the fly wheel of the engine by means of a spring-loaded pressure plate. When the pedal of the clutch is depressed, the pressure plate is pushed back by the release levers. This releases the pressure from the clutch plate. Then the clutch plate stops rotating, but the flywheel continues to rotate. When the clutch pedal is released, the pressure plate forces the clutch plate against the flywheel, causing the clutch plate and the flywheel to turn together as one unit. Thus, the power of the engine goes to the gear box for onward transmission to the rear wheels.

Parts of a single clutch and their functions are given below:

Parts	Description
Cover plate	Covers the whole assembly
Pressure plate	With the help of pressure spring, it exerts pressure on the clutch plate
Clutch plate (12")	Transmits motion during its engagement with the fly wheel
Pressure spring	Exerts pressure on the pressure plate which in turn exerts pressure on the clutch plate
Insulator	Prevents conduction of heat generated during working to the spring
Release lever	Acts as a fulcrum, together with link and adjusting screw
Adjusting screw	Acts as a fulcrum, together with release lever and link
Antirattle spring	Gives cushioning effect on the finger or release levers, when the clutch pedal is released

(ii) Multiplate clutch

Function: Transmits power from the engine to the transmission through a set of clutch assembly (primary) and to the hydraulics through the other clutch assembly (secondary). These clutches are located together one behind the other. Partial depression of the clutch pedal (halfway down) cuts off the power supply only to the transmission, while the hydraulics will receive the power. Full depression of the clutch pedal cuts off the power supply to both transmission and hydraulics.

Parts of a multiplate clutch and their function are given below:

Parts	Description
Cover plate	Covers the whole assembly
Beleville spring	Exerts pressure on the secondary pressure plate which in turn exerts pressure on the secondary clutch plate
Secondary pressure plate (PTO)	With the help of a Belleville spring, it exerts pressure on the clutch plate
Secondary clutch plate (10")	Transmits the motion to the hydraulic pump during its engagement with the secondary flywheel
Secondary fly wheel	Power from primary fly wheel of engine is transmitted to the secondary fly wheel which operates the hydraulic pump and PTO
Primary pressure plate	With the help of a pressure spring, it exerts pressure on the clutch plate
Clutch plate (12")	Transmits motion during its engagement with the flywheel
Pressure spring	Exerts pressure on the pressure plate which in turn exerts pressure on the clutch plate

Insulator	Prevents conduction of heat generated during working to the spring
Release lever	Acts as a fulcrum together with the link and adjusting screw
Link	Acts as a fulcrum together with the release lever and adjusting screw
Adjusting screw	Acts as a fulcrum together with the release lever and link
Antirattle spring	Gives a cushioning effect on the finger or release levers when the clutch pedal is released

(iii) Diaphragm clutch

The diaphragm clutch transmits more torque as the diaphragm exerts more force as compared to the coil springs.

Advantages

Light weight

Enables smooth shifting of gear

Exerts consistent pressure plate load due to diaphragm spring design

Long lifespan

Easy handling in tough soil conditions

(iv) Split torque clutch

Operates the tractor's transmission drive using a single pedal release mechanism whilst utilizing the cover to power the PTO (implement drive).

Design concept

The terms "Split Torque" and "Fixed PTO" are often used to describe the operation of this clutch. The cover has an internally fixed splined hub providing power to the hydraulic PTO pack. This is operated independently by means of a lever located within the cabin. The plate clamping method is usually a single diaphragm spring or coil spring and lever type arrangement; the cover is used in conjunction with a single clutch disc mounted against the flywheel.

What is split torque transmission?

Split torque transmission is a form of automatic motor vehicle transmission which is able to send power to two separate places at the same time, and even control how much power is sent to each output source. Such transmissions were first invented for use in helicopters, to control both main and tail rotors at once. However they have become popular with all-wheel drive vehicles as well. New models incorporate an onboard computer that senses how much power is needed in each drive axle to maintain maximum traction and adjusts the power distribution to compensate.

Structure

The layout of split torque transmission has a few key differences from normal automatic transmission. Whereas normal transmission has a single set of planetary gears connected to the engine by way of the flywheel, split torque transmission has two such gear sets sitting side by side. There are a series of clutches sitting between the flywheel and each of the gear sets (Fig 3.1.1.). These clutches are controlled automatically by a series of valves which open and close in relation to the position of a self-orienting electromagnetic solenoid. Two crankshafts emerge from the transmission, one for each gear set. Each is connected to a different drive axle, meaning one powers the front tyres and the other powers the rear tyres of a vehicle.

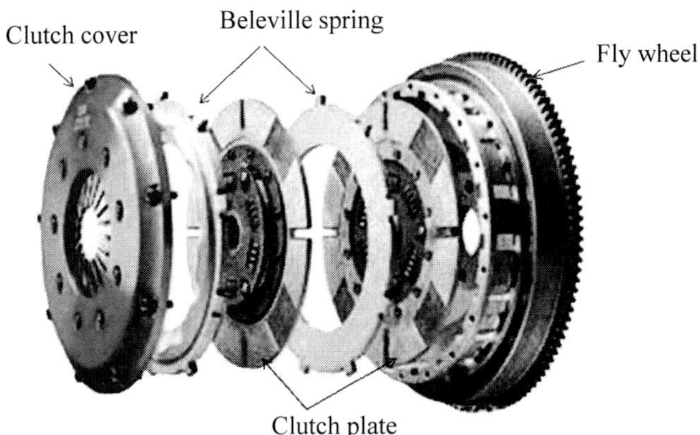

Fig 3.1.1. Split torque clutch
Source for fig: www.rapid-racer.com

How does split torque transmission work?

Starting with first gear, the engine will send power equally to both the planetary gear sets. As oil pressure and engine performance increase, the valve bodies controlling the clutches and brakes attached at key points on the planetary gear set will engage, causing the planetary gear set configurations to change, bringing the vehicle into second gear. The same process repeats whenever a gear change occurs. On a flat terrain while the vehicle moves straight, the flywheel sends equal amounts of power to both gear sets, and thus an equal amount of power reaches all the tyres. The electromagnetic solenoid changes orientation to stay upright when the vehicle tilts, either as a result of a hard turn or while going up and down on an inclined road. The solenoid having seated differently in its housing sends a signal to the valves connected to the clutches on the flywheel. Depending on the solenoid's position, some valves open while the others close. This causes some clutches to engage while others to release. This changes the distribution of power going from the flywheel to the gear sets.

3.1.5. Servicing of the clutch

The clutch is subjected to wear and tear and requires adjustment. Two types of adjustments can be done. One is the simple free play adjustment that can be done by any operator in consultation with a mechanic. The other is clutch release lever adjustment that is done at the dealership. Clutch lever adjustment requires special tools and fixtures that are available only with an authorized dealer.

3.1.6. Trouble shooting

These are the troubles faced by owners of tractor of any make or model

Fault	Probable causes
Clutch will not release	Oil or grease on friction plate. Damaged pressure plate on clutch cover. Improper pedal adjustment. Friction plate hub binding on splined drive pinion. Distorted friction plate. Broken facings on friction plate. Dirt or foreign matter in the clutch
Clutch slip	Oil or grease on friction plate. Weak or soft pressure springs. Binding of clutch pedal mechanism preventing its full return to stop. Improper pedal adjustments preventing full engagement
Juddering clutch (Tractor moves in a series of jerks when the clutch pedal is released)	Loose or worn out clutch. Distorted clutch plate. Misalignment of pressure plate with the flywheel. Bent input shaft. Oil, grease and dirt in the friction plate
Dragging clutch (When the operator presses the pedal to disengage the clutch, the driven plate will not stop rotating)	Too much pedal free play. Damaged pressure plate. Cracked clutch-driven plate. Clutch-driven plate seized in input shaft splines. More run out in the driven plate.
Rapid wear of lining	Operator rides on the clutch pedal. Weak splines on the input shaft and driven plate
Clutch noise	Flywheel not tightened properly. Improper adjustment of clutch linkage. Worn out clutch release bearing. End float in rear axle.

3.2

DriveTrain or Transmission

3.2.1. Gear boxes

It is a unit fitted after the clutch. Forms a part of transmission and provides speed and torque conversions from a rotating power source to another device using gear ratios. Its primary function is to change the speed of the tractor to cater to varying field, road and load condition. It also helps in reversing the tractor.

Gears are usually made of alloy steel. Since the tractor has to transmit heavy torque all the time, the best quality lubricants, free from sediments, grit, alkali and moisture, are used for lubrication. SAE 90 oil is generally recommended for gear box.

The description about gear boxes stated below pertains to major tractors manufactured in India. The number of teeth in each of the gear and the engagement between gears of the main shaft and the corresponding gear on the counter/lay shaft also change in different tractors evolved over a period of time. Hence, information hosted in this section is only indicative.

3.2.2. Types

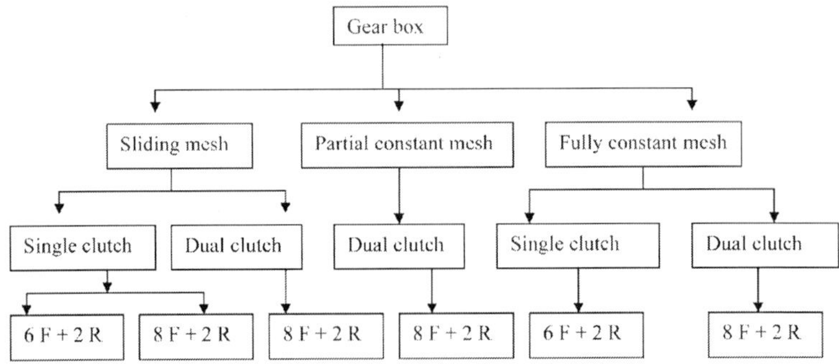

F: Forward, R: Reverse

3.2.3. Common construction

a. Main drive shaft

The main drive/input shaft is located in the housing found in the front compartment of the transmission case. The main/input shaft is mounted on a ball bearing and carries an integral helical gear in case of 6 forward and 2 reverse single sliding mesh and an integral spur gear in case of 8 forward and 2 reverse single/dual sliding mesh gear box. It will be in constant touch with the gear located on the counter/lay shaft.

b. Counter/lay shaft

The counter/lay shaft is supported on two ball races located in the centre web and rear wall of the transmission case. The countershaft is solid in the single clutch type (6F + 2R, 8F + 2 R) and hollow in the dual clutch type (8 F + 2 R). It has 15 gear teeth machined on its outer diameter which serves as the first gear.

Mounted on the lay shaft are two gears (6F + 2 R) and three gears (8 F + 2 R- both single and dual clutch types), which are 3rd and 2nd and 4th, 3rd and 2nd gear pinions, respectively. None of the lay shaft gears are free to move along the shaft, either being retained by abutment with other gears or bearings or snap rings.

c. Main shaft

The main shaft has externally splined gear teeth. In 8 F + 2R dual clutch type, it has a bore at its front end to accept the spigot of the main drive shaft and its caged roller bearing.

Mounted on the main shaft are:

a. Three gears: First, third and second in case of 6 F + 2 R. The 3rd and 2nd gears are compound gears named cluster gears

b. Four gears: Fourth, first, third and second in case of 8 F + 2 R. The 3rd and 2nd gears are compound gears named cluster gears

The main shaft gears are free to slide with the help of selector forks, which are externally connected by a gear shift lever.

d. Epicyclic unit

The basic 3 forward and one reverse and 4 forward and one reverse gear (single and dual clutch types) are doubled by the epicyclic unit mounted at the rear end of the transmission case. The epicyclic unit comprises a ring gear, inside which runs three planetary pinions mounted in a carrier. Three planetary pinions with gear teeth are positioned equidistant on the periphery of the main shaft, which acts as the sun gear. When the main shaft (sun gear) rotates, the planetary pinions also rotate, but being meshed with the teeth on the inside of the ring gear, they revolve at a lower speed. The rotational speed of the carrier is reduced by a rotation of 4:1.

To transmit the drive from the epicyclic unit to the rear axle, a drive shaft is connected by a coupler, either directly to the gearbox main shaft (sun gear-high range) or to the planetary pinion carrier (low range). Movement of the dual range selector lever actuates the high–low rail. This high–low rail in turn moves the selector fork and the coupler. The coupler either meshes with the main shaft (high) or to the planetary pinion carrier (low). Between the two engaged (high or low range) positions, there is a neutral position. The neutral position can be achieved by keeping the coupler in between the main shaft and planetary pinion carrier. A safety neutral switch is fitted on top of the high–low rail at its front end. This switch provides continuity of power supply to the engine starter motor only when the dual range selector is in neutral.

e. Reverse gear cluster

The reverse gear cluster is a single piece of construction fixed on a shaft and is located in the transmission case. This consists of an idler gear with 13 teeth in one half and 21 teeth in the other half. The 3rd gear in the lay shaft is in constant mesh with the 21 teeth of the reverse gear cluster. When the reverse gear is selected, the 1st/ reverse gear on the main shaft comes into contact with 13 teeth of the reverse gear cluster. The engine power is now transmitted through the main drive shaft to the lay shaft and to the reverse gear cluster (21 teeth).

The drive is further transmitted by the reverse gear cluster (13 teeth) to the main shaft 1st/reverse gear. The drive transmitted from the engine is thus reversed in direction.

f. Lubrication

The entire transmission case is bathed in lubricating oil. This prevents metal contact. In the absence of lubricant, a lot of heat will be generated resulting in burning of gear teeth. The oil also lubricates the bearings. A magnetic drain plug is fitted in the transmission case to attract small iron particles which usually get mixed in the lubricating oil when newly assembled gears mesh with each other. These iron particles are removed whenever the lubricating oil is changed.

g. Gear levers

There are two gear levers, *viz*., (a) the gear shift lever for the change of speed, in which the three/four forward gear and one reverse gear positions are indicated and (b) the dual range selector lever meant for the reduction unit, in which the dual range selector has high, neutral and low range positions indicated on the knob.

The dual range selector lever must be in its start (neutral) position to enable the starter motor circuit with the safety neutral switch. This allows the tractor to be started. A low or high range must be selected before the tractor is moved. When a low range is engaged, the reduction unit is also engaged and effects a 4:1 reduction of all transmission speeds (Fig 3.2.1).

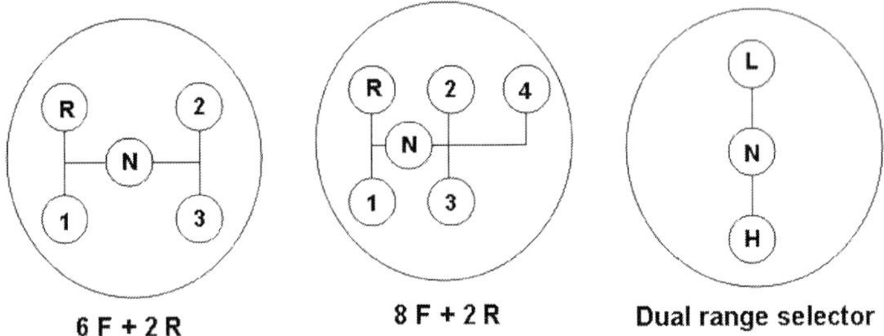

Fig 3.2.1. Gear levers

h. Gear ratio

When two gears have the same number of teeth, then they both run at the same speed. In this case, the gear ratio is 1:1. When a smaller gear 'A' having 6 teeth drives a bigger gear 'B' having 18 teeth, the smaller gear has to run 3 times to make the bigger gear 'B' undergo one revolution. In this case, the gear ratio will be 18:6 or 3:1 (Fig 3.2.2). Thus, the gear ratio of two gears is their relative speed or the rpm on which they run.

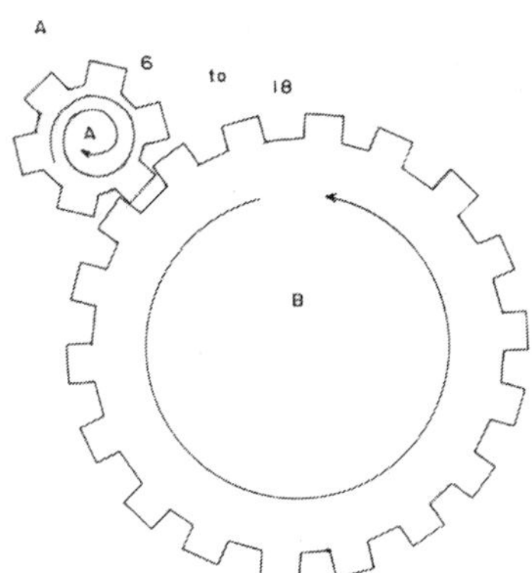

Fig 3.2.2. Gear ratio

i. Details of gears

Items	Single		Dual
	6 F + 2 R	8 F + 2 R	8 F + 2 R
Number of forward speed	6	8	8
Number of reverse speed	2	2	2
Reduction unit ratio	4: 1	4: 1	4: 1
PTO engine speed reduction	2.78:1	2.78:1	3.31:1
Oil capacity (litres)	30.28	30.28	30.28

j. Speed reduction ratio

Gears	Single	Single	Dual
	6 F + 2 R	8 F + 2 R	8 F + 2 R
Low			
L_1	205.5:1	205.5:1	200:1
L_2	137:1	137:1	136:1
L_3	74.7:1	74.7:1	74.3:1
L_4	NA	60.9:1	60.5:1
Reverse (L)	150.9:1	150.9:1	148:1
High			
H_1	51.4:1	51.4:1	49.9:1
H_2	34.2:1	34.2:1	34:1
H_3	18.7:1	18.7:1	18.9:1
H_4	NA	15.3:1	15.5:1
Reverse (H)	37.73:1	37.7:1	37:1

NA- Not applicable

k. PTO drive shaft

The clutch plate drives the input constant mesh gear through the main drive shaft. This in turn drives the solid countershaft and the hydraulic pump shaft. The hydraulic pump shaft in turn drives the PTO shaft through the PTO coupler.

3.2.4. Unique features and differences of different gear boxes

a. Sliding mesh gear box

The most commonly used gear box in tractors is the sliding mesh type. In this type, gears of the main shaft are slided to mesh with their corresponding pairs in the countershaft. This achieves the required torque or speed.

Types of sliding mesh: The basic type is driven by a single clutch with variants like 6 forward and 2 reverse speeds (6 F + 2 R) and 8 forward and 2 reverse speeds (8F + 2R). A combination of three forward and one reverse gear and its output when subjected to one more reduction through epicyclic gear train add up to the availability of 6 F and 2 R speed. In the same manner, four forward and one reverse gear result in 8 F + 2 R speed gear box. A brief description in the subsequent paragraphs would explain how the gear arrangements are worked out.

The basic design of all the gear boxes is similar. The main drive shaft (input shaft) takes the drive from the flywheel of the engine through the clutch. The main drive shaft in turn rotates the countershaft (lay shaft). The countershaft again rotates the main shaft, which in turn drives the rear wheels through the rear drive shaft, pinion, crown and differential and rear axle.

Single clutch: 6 F + 2 R

The input shaft with 18-teeth helical integral gear is driven by the main transmission clutch plate and drives a constant mesh gear (50 teeth). The 50-gear teeth in turn drives a lay shaft to which is attached the 2nd and 3rd gears. The first gear is integral on the lay shaft. The main shaft is located above the lay shaft and three sliding gears are mounted on it, one of which is a compound gear, having 2 sets of teeth.

The sliding gears are operated by selected forks mounted on shifter rails which in turn are operated by the gear selector lever selected located directly above the gearbox.

The sliding gears mesh with the lay shaft gear drive from the front, 1st, 3rd and 2nd gears, respectively, depending on the gear selected by the lever (Fig 3.2.3).

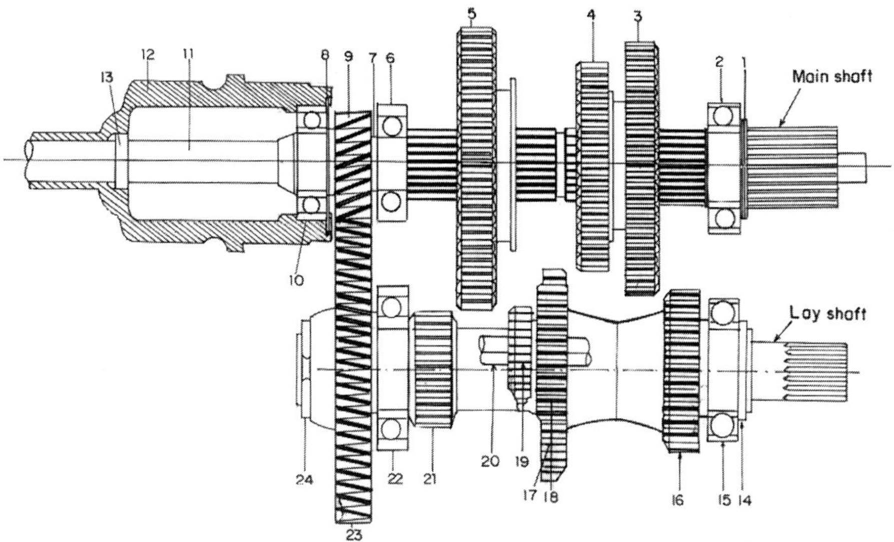

Fig 3.2.3. Sliding mesh - Single clutch - 6 F + 2 R

1,7,8,14,24- Snap ring 2,6,10,15,22- Ball bearing 4,17- Third gear 3,16- Second gear 5,21- First gear 11- Main drive shaft 12- Main drive shaft housing 13- Oil seal 18,19- Gear reverse cluster 20- Shaft reverse cluster 9,23- Gear constant mesh

Single clutch: 8 F + 2 R

The operation is similar to that described for 6 F + 2 R. But this has an additional pair of gear to cater to the fourth gear combination.

Dual clutch: 8 F + 2 R

The dual clutch gear box came into existence to incorporate the following additional features not found in single clutch type gear box.

(a) The hydraulic pump or the hydraulic operation keeps working even when the primary clutch is pressed or disengaged to bring the tractor to the halt position

(b) PTO-driven implements keep working even when the primary clutch is disengaged

The dual clutch drives two main drive shafts with integral gear (one shaft is solid and is for the transmission with 17 teeth gear, while the other is for PTO and is hollow with 16 teeth gear). The main transmission clutch plate drives a pair of constant mesh gears (17 and 52 teeth). The 52 gear teeth in turn drives a hollow lay shaft to which is attached the 4th, 3rd and 2nd ratio gears. The first gear is integral on countershaft and has 15 teeth. The main shaft is located above the lay shaft and three sliding gears are mounted on it, one of which is a compound gear having 2 sets of teeth. The PTO clutch plate drives a pair of constant mesh gears (16 teeth and 53 teeth). The 53 teeth gear drives the hydraulic pump and the PTO shaft.

b. Partial constant mesh

The reason for the development of partial constant mesh is to make a tractor driver change between the first and reverse gear with ease when ploughing in wet lands characterized by small plots with insufficient head space and heavy clayey soil. The driver can go forward and reverse the tractor in the same direction, simultaneously using the first and reverse gear. This type of design reduces the wear and tear of gears and facilitates easy manoeuvring.

The partial constant mesh gear box has independent 44 teeth gear each for the 1st and reverse gear on the main shaft. It has only the first gear and reverse gear of the main shaft in constant mesh with the respective gears of the countershaft. The rest of the gears are of sliding type as found in the sliding mesh gear box. The splined bushes, fixed sleeve and shift sleeve are fitted between the first and reverse gear constant mesh pairs. The first and reverse gear are engaged by sliding the shift sleeve either to mesh with the first gear of the main shaft or to the reverse gear of the main shaft, respectively.

c. Fully constant mesh gear box

In the fully constant speed gear box, the gears of the main shaft and countershaft are always in the engaged (constant mesh) position. The gears of the main shaft are mounted over splined bushes and are free to rotate on it. The drive from the input shaft is fed to the countershaft, which in turn rotates all the gears of the main shaft even when the transmission is in neutral as they are in constant mesh.

If A, B, C and D is the forward and E is the reverse gear on the main shaft, then, the corresponding A1, B 1, C1, D1 and E1 found on the lay shaft mesh with each other. Between gears on the main shaft (A and B and C and D), a shift sleeve is placed over the fixed sleeve (Fig 3.2.4). During selection of gear, the shift sleeve slides and engages with the respective gear. Only then, the drive gets transmitted to the main shaft. When the reverse gear is selected, the shift sleeve engages with the reverse gear of the main shaft. The fixed sleeve has both internal and external splines. The internal splines mesh with the main shaft, while the external spline meshes with the shift sleeve. The shift sleeve has internal splines meshed with the fixed sleeve, while the outer has provision for the shifter fork. The transmission path is as follows:

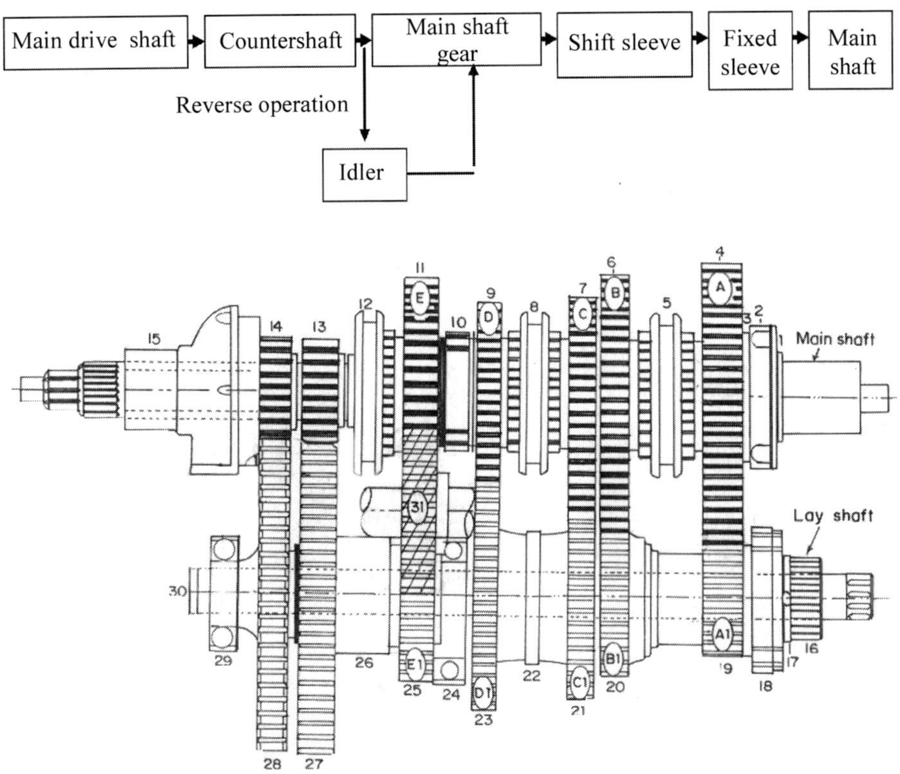

Fig 3.2.4. Fully constant mesh – Dual clutch - 8 F + 2 R

1,17- Snap ring 2- Ball bearing 3- Thrust washer 4- First gear 5,8,12- Shift sleeve 6- Second gear 7- Third gear 9- Fourth gear 10- Roller bearing 11- Reverse gear 13- Input pinion 14- PTO pinion 15- Main drive housing 16- Counter shaft 18- Roller bearing 19- Pinion first gear 20- Pinion second gear 21- Pinion third gear 22- Spacer composite 23- Pinion fourth gear 24- Ball bearing 25- Reverse gear 26- Spacer 27- Transmission gear 28- PTO gear 29- Bearing 30- PTO drive shaft 31- Reverse idler gear

d. Gear arrangement in high hp tractors

Gear boxes in high hp tractors (50-60) of different manufacturers have either partial constant mesh or fully constant mesh or partial synchromesh or forward–reverse shuttle with full synchromesh types with speeds of 8 forward + 4 reverse, 9 forward + 3 reverse, 8 forward + 8 reverse or 12 forward + 12 reverse. A few of these are also fitted with differential lock. The different types of gear boxes and their speed combination are available in the manufacturer's catalogue.

3.2.5. Differential unit and final drive

a. Differential

A differential unit is a special arrangement of gears that permits one of the rear wheels of the tractor to rotate slower or faster than the other. While turning the tractor on a curved path, the inner wheel has to travel a lesser distance than the outer wheel. The inner wheel requires lesser power than the outer wheel. This condition is fulfilled by the differential unit, which permits one of the rear wheels of the tractor to move faster than the other at the turning point. The output from the gear box is transferred to the rear wheel with the pinion and crown. The main functions of the pinion and crown are to:

(i) transmit power through a right angle drive to suit the tractor wheels.

(ii) reduce the speed of rotation.

The differential is fixed on the right hand side of the crown wheel. The axles are driven through the differential.

The differential unit consists of (a) differential casing, (b) star gears on spider or cross, (c) sun gears, and (d) rear axle shaft.

When the tractor moves in a straight line, the star gear (usually four numbers) inside the differential case does not rotate about its axis, but gets locked with the sun gears (two numbers). The star gears get meshed and drive the sun gears which in turn rotate the two axle shafts. Thus, the drive is imparted equally to both the wheels. When the tractor takes a turn the inner wheel will rotate at a speed lower than its outer wheel. This lower speed of the inner wheel is transmitted through the axle shaft to the inner sun gear. But the differential case along with the crown wheel is still driven by the pinion. The four star pinion gears inside the differential case rotate about their own axis. This in turn drives the axle shaft and outer rear wheel faster. This action of the differential allows the tractor to negotiate bends smoothly without losing stability. By braking one side wheel independent of the other, a sharp turn can be achieved. This is found useful for turning the tractor in small fields.

b. Differential lock

Differential lock is a device that joins both half axles of the tractor and makes them into a single unit, so that even if one wheel is under less resistance, the tractor comes out from the mud, etc, as both wheels move with the same speed and apply equal traction.

c. Final drive

The final drive is a gear reduction unit achieved either through planetary drive or by bull gear. This drive transmits the power finally to the rear axle and the wheels. The tractor rear wheels are not directly attached to the half shafts, but the drive is taken through a pair of spur gears. Each half shaft terminates in a small gear which meshes with a large gear called bull gear. The bull gear is mounted on the shaft, carrying the tractor rear wheel. The device for final speed reduction, suitable for tractor rear wheels, is known as the final drive mechanism.

3.2.6. Trouble shooting

An exhaustive list of troubles faced by tractor owners of any make pertaining to gears is provided below. Mechanics with sufficient experience in the authorized dealership can diagnose the problem.

Fault	Probable causes
Gears not engaging properly	Free play of clutch pedal. Clutch adjustment. Plunger spring. Gear shift lever. Shifter rail grooves. Rail shift bore. Main shaft bearing. Countershaft bearing.
Noisy gear box	Less or no lubricating oil in gear box. Gear box or flywheel housing out of alignment with the engine. Worn out or broken gear teeth. Loose gear on splines or shaft. Worn out teeth of shifting sleeve. Worn out bearings of clutch shaft or main shaft or countershaft. Bent fork. Worn out gear shifting mechanism. More/less backlash in gears
Slipping out of gear	Weak spring of rail shifter. Engine flywheel housing and gear box out of alignment. Worn out gear of sleeve or gears. Worn out pilot bearing in clutch shaft. Worn out bearing of clutch shaft, main shaft or countershaft. Worn out or bent fork. Worn out rings of synchro mesh unit. Too much play in gear shifting mechanism. Wrong selector ball design
Oil leak	Too much of oil in the gear box. Oil of low viscosity. Loose top cover or any other cover bolts. Cracked housing or top cover. Worn out rear oil seal. Broken gasket. Worn out sleeve of release bearing. Loose drain plug or filler plug.
Noisy differential	Less lubricating oil in the differential housing. Low viscosity oil or use of poor quality oil. Wrong adjustment of crown wheel and pinion. Worn out or broken teeth of crown wheel or pinion. Less back lash of crown wheel and pinion teeth. Worn out bearing of crown wheel or pinion. Crown wheel misaligned on cage. Loose star pinion. Broken or worn out thrust washers of star pinion or sun pinion
Tractor does not move when put in gear	Broken axle shaft. Broken teeth of crown and pinion. Broken cross or star pinion. Broken coupler rear drive/shear tube. Worn out or glazed friction liner (clutch slipping). Clutch hub spline shear.

3.3

Rear Transmission

3.3.1. Main components

Shear tube: It absorbs heavy shocks and protects the components of rear transmission and main transmission from breaking.

Crown wheel and pinion: It transmits the drive from the main shaft to the rear axle shaft and inverts the direction of rotation to 90° from the main shaft. It reduces the speed and multiplies the torque.

Rear axle: It is a semi floating type. It transfers the drive from the sun gear to the rear wheel.

3.3.2. How does it work?

The drive from the main shaft of the gear box is taken to the rear drive shaft through the coupler planetary. Depending on the position of the high–low gear and lever which moves the coupler, power is transmitted directly or through epicyclic gear train. The drive from the rear drive shaft is transferred to a safety coupling also known as shear tube. The coupling is a safety device. In case the rear wheels encounter any shock loading, the coupling breaks and disconnects the gear box from the rear transmission. This prevents damage to costly parts in the transmission, clutch and engine. Power is transmitted from safety coupling to a pinion which has a spiral bevel gear at one end. The bevel pinion is supported on two tapered roller bearings carried in a housing which is mounted in the centre housing.

The rear end of the bevel pinion shaft is supported on a roller bearing, so that it does not deflect. The bevel gear drives a spiral crown wheel. The crown wheel is riveted/bolted to a differential case which may be in two or three pieces. On both sides of the differential case, taper roller bearings are fixed, which in turn are supported on bearing cups on the left and right side axle housings. The differential case assembly has a cross and on each arm carries a pinion gear. These are called planetary gears. Thrust washers are provided at the back of each pinion gear to absorb the thrust encountered by them. Two bigger size bevel gears are placed inside the differential case which is always in a mesh with four planetary gears. These bigger gears are called sun gears or differential gears and have internal splines. Non-rotating thrust washers are also fitted behind these gears. Two semi floating axles are supported by the internal splines of the gear differential at the inner end and by a tapered roller bearing housed in a retainer bearing housing at outer end.

The two rear axle shafts get the drive through the two differential gears. The left hand side axle housing has a thrust pad fitted to take care of any deflection of the crown wheel due to sudden load. The shrunk fit collar on both the axles prevents it from coming out of the retainer housing (Fig 3.3.1)

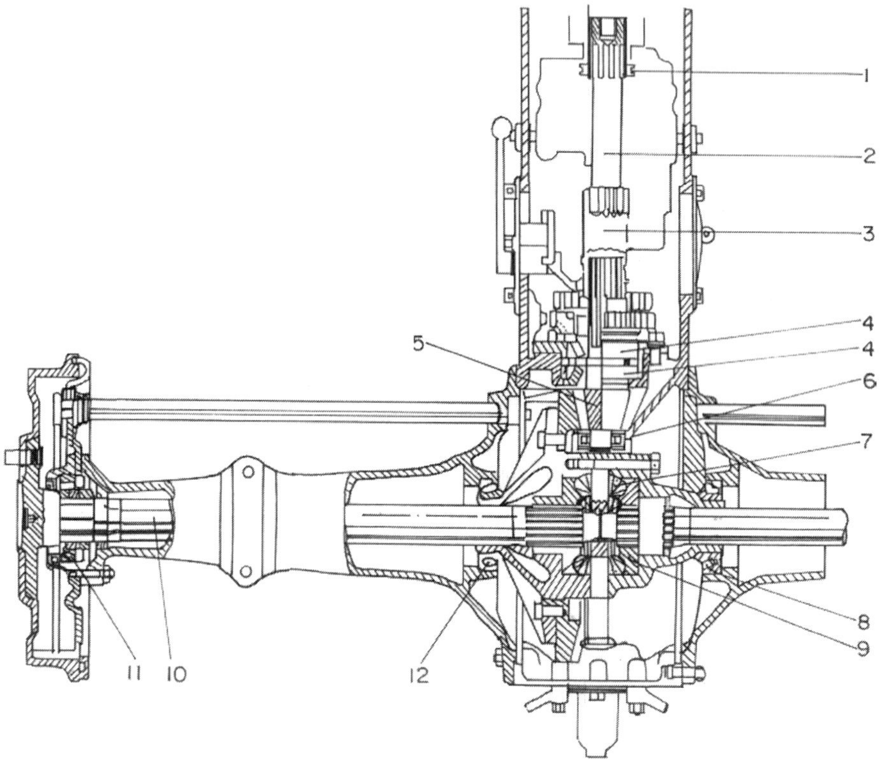

Fig 3.3.1. Rear transmission

1- Coupler 2- Rear drive shaft 3- Shear tube 4,8,11,12- Taper roller bearing 5-Pinion 6- Pilot bearing 7- Planetary gear 9- Differential gear 10- Axle shaft

3.3.3. Faults in the rear axle and probable causes

These are the troubles faced by tractor owners of any of make pertaining to rear transmission.

Fault	Probable causes
Humming nose (in motion)	Worn out rear axle bearings
	Crown wheel and pinion backlash
	Worn out crown wheel and pinion bearings
	Crown wheel and pinion damage
	Worn out crown wheel and pinion /differential thrust washer
	Rear axle end float not adjusted properly
Other noises (grinding, hitting, etc.)	Oil level condition
	Worn out rear axle bearings
	Rear axle end float not adjusted properly
	Incorrect pinion preload
	Worn out pinion bearings
	Worn out pilot bearings
	Crown wheel and pinion damage
	Differential gear/differential case damage
	Worn out thrust washer/ differential case bushes
	Worn out thrust pad.

3.4

Brakes

3.4.1. Function

The brake is used to stop or slow down the motion of a tractor. It is mounted on the driving axle and operated by two independent pedals. Each pedal can be operated independently to assist turning of the tractor during field work or both the pedals can be locked together.

3.4.2. Principle of operation

The brake works on the principle of friction. When a moving element is brought into contact with a stationary element, the motion of the moving element is affected. This is due to frictional force which acts in the opposite direction of the motion and converts the kinetic energy into heat energy.

3.4.3. Classification of brake

a. Mechanical brake (Fig 3.4.1.)

Internal expanding shoe type

Two brake shoes made of frictional material are fitted inside the brake drum. These are held away from the drum by means of springs. One end of each shoe is fulcrumed, whereas the other is free to move by the action of a cam which in turn applies force on the shoes. The movement of the cam is enabled by the brake pedal through the linkage. The drum is mounted on the rear axle, whereas the shoe assembly is stationary and is mounted on the back plate. Internally expanding 14" x 2" sizes are usually fitted in many tractors with bonded linings in production, but the shoes are drilled to receive riveted linings in service.

Between the shoe webs at the anchor pin end is the operating camshaft, which is connected by a linkage to the pedal. The shoes are kept square in relation to the back plate by steady posts and the shoe holds down the pins. The back plate is secured to the rear axle housing and is enclosed within a drum, which is fitted to the rear axle shaft assembly. This is also known as drum brake.

Operation

Brakes are operated by two independent brake pedals situated on the right side of the transmission case. Each pedal can be operated independently to assist turning during field work or locked together by means of the combining brake lock pivoting on the right hand pedal for on road operation.

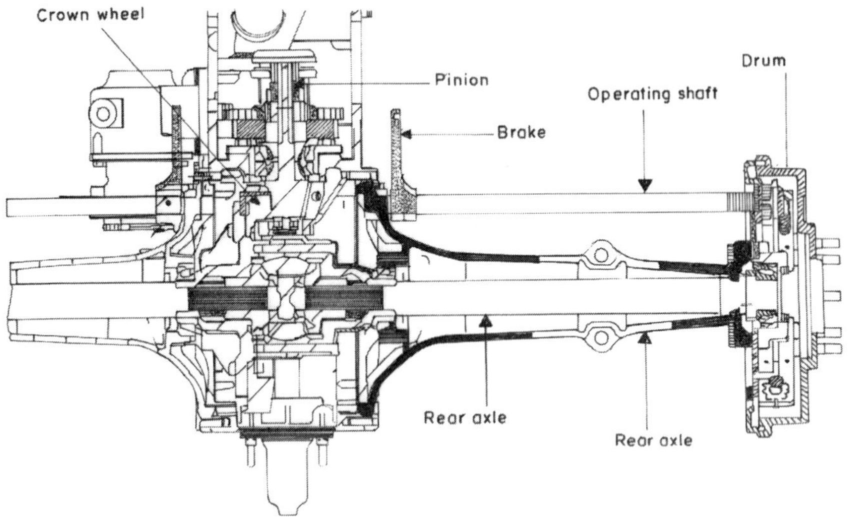

Fig 3.4.1. Mechanical brake system

External contracting shoe type

This type of brake system is normally available on crawler tractors. The drum mounted on the drive axle is directly surrounded by the brake band. When the pedal is depressed, the band tightens the drum.

Disc brake

Two actuating discs have holes drilled in each disc in which steel balls are placed. When the brake pedal is depressed, the links help to move the two discs in opposite directions. This brings the steel balls to shallow parts of the holes drilled in the disc. As a result, the two discs are expanded and the braking discs are pressed between the discs and the stationary housing. The braking discs are directly mounted on the differential shaft which ultimately transfers the travelling effect to the differential shaft. Operation of different types of disc brake in the tractors is given below.

Sealed dry disc brake (Fig 3.4.2.)

Operation

- When force is applied on the brake pedal, this results in the actuating assembly to come into contact with two rotating friction discs splined on to each axle shaft through the hub. These in turn come into contact with the fixed friction faces provided in the brake housing and in the end cover.
- The mechanism of each brake consists of two cast iron actuating discs, held together by tension springs and separated by steel balls through push rods relative to the other. The steel balls ride up their inclined seats and so spread the actuating disc apart.
- The actuating disc comes into contact with the rotating friction discs. The torque of one actuating disc comes into contact with lug in the brake housing.
- The other actuating disc tends to rotate further, increasing the angular displacement between the discs, and assists in braking
- When the operating pull is released, the tensioned springs cause the discs to return to their normal position.

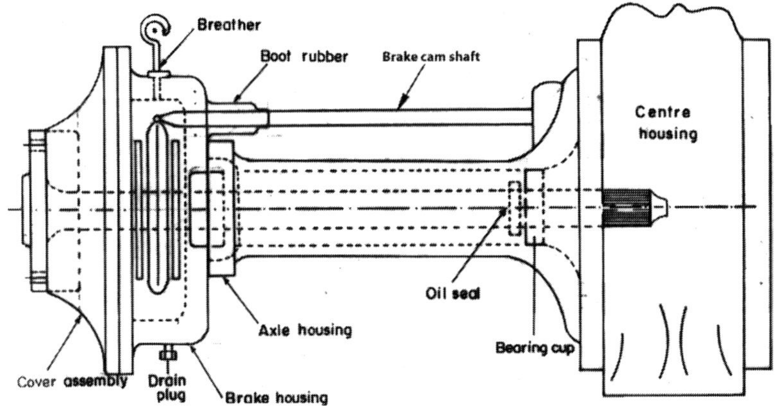

Fig 3.4.2. Sealed dry disc brake system

Oil- immersed/Multiple disc wet type brake (Fig 3.4.3.)

Operation

- The foot brake system is mechanically operated.
- The two pedals can be used independently to aid turning in confined spaces or can be latched together to provide a master pedal for normal braking of both rear wheels.
- The brake pedals must be locked together to ensure uniform brake application and maximum stopping of brakes.
- The brake is unlocked when a sharp turn is made. Sharp turns, however, must be made only at slow speeds. The inner pedal will then act on the left hand wheel.

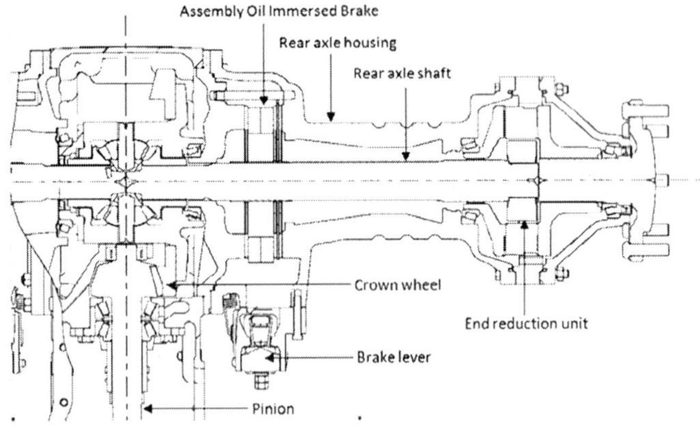

Fig 3.4.3. Oil-immersed/Multiple disc wet type brake system

b. Hydraulic brake

The hydraulic brake system is based on the principle of Pascal's law. The brake fluid which is usually a mixture of glycerin and alcohol is filled in the master cylinder; when the pedal is depressed, the piston of the master cylinder is forced into the cylinder and the entire system turns to a pressure system. Immediately, the piston of the wheel cylinder slides outward which moves the brake shoes to stop the rotating drum. When the pedal is released, the return spring of the master cylinder moves the piston back to its original piston, causing a sudden pressure drop in the line. The retracting springs of the brake shoe bring them back to their original position. Thus, the piston of the wheel cylinder returns.

c. Parking brake

- The parking brake acts on the rear wheel.
- To engage the brake, the foot brake pedals have to be pressed and the hand lever is pulled up.
- To release the parking brake, the foot brake pedal is pressed. The button at the end of the hand lever is pressed and the lever pushed down.

3.4.4. Adjustments

a. Drum brake (Mechanical internal expanding)

A jack is used to clear rear wheels off the ground. It is ensured that all shafts and pins work freely. The inspection cover plate is moved to one side. A screwdriver is inserted through a slot and the clicker adjuster is levered (towards the centre of the tractor) to expand the shoes in the drum until the wheel is locked. The adjuster is then slackened off until the wheel is free to rotate. The above adjustments are done for both the rear wheels. To test the brakes, fourth gear is engaged whilst driving at a slow speed. With brake pedals locked together, the brake is applied firmly. Any tendency to veer off course is counteracted by slackening the adjuster on the side towards which veering takes place. It is always ensured that the brakes are maintained properly for efficient operation. It is a must that brake shoes and drums are cleaned thoroughly after a puddling operation. The brakes are checked every 200 hrs, or more frequently if heavy work is involved.

b. Sealed dry disc brake

- The tractor is jacked until both rear wheels are clear off the ground. Then the brake is jammed with a screw adjuster completely on both sides. The horizontal brake rod is connected with the brake lever. It is also ensured that there is an equal thread engagement on both ends of the clevis. The hand brake linkages are then disconnected. The pedal play is arrested to zero by tightening the horizontal brake rod. The pedal free play is kept at 70 mm by adjustment. This is done by keeping the left pedal in the home position as a reference for setting the right pedal. In the same way, the process is repeated for the right pedal. The lock nuts are then tightened in the horizontal brake rods on both the sides. The screw adjuster is released by one full revolution (7 flats). The lock nuts are then tightened. The overall free play is checked for travel length. It should be around 85 mm.

- During brake trials, if there is uneven braking between the right and left sides, the adjustment is carried out by loosening the screw adjuster on the brake housing.
- The hand brake linkages are connected after full adjustment to ensure that free play of at least one notch, *i.e.* the brake pedal movement is not present when the hand brake lever is moved through one notch. It is also ensured that during application of hand brakes, the right and left pedals get depressed to the same height in the unlatched condition.
- Regular adjustments (wear adjustments) are carried out by jamming the brakes through the screw adjuster and loosening the screw adjuster by 7 flats. No adjustments on linkage are carried out.

c. Oil-immersed/Multiple disc wet type brake

The tractor is jacked up until both rear wheels are clear off the ground. The brake pedals are then unlatched. Parking brake linkage in the brake pedal is then removed on both the sides. The wheel is rotated and the brake adjuster nut is tightened. The nut is removed when the rotation of the tyre comes to a halt. A wedge of 145-150 mm is provided below the pedal and foot rest. The linkage is then fixed in proper alignment and the lock nut adjusted if necessary. The adjuster nut is then loosened by rotating one revolution. The same procedure is adopted for right hand side.

3.4.5. Trouble shooting

These are the troubles faced by tractor owners of any make pertaining to brakes

Fault	Probable causes
Brake lock (brake remains applied without pressing the pedal)	Foreign material. Incorrect lubrication. Wrong adjustment of pedal linkage
Brakes do not work	Worn out brake linkages. Broken/disconnected pedal linkages
	Blocked breather resulting in discs soaked in oil (applicable to dry disc brake)
Spongy brake	Presence of air in the system because of less fluid in the master cylinder reservoir. Faulty check valve
Noisy brake	Loose rivets or worn out lining will cause chattering. Rough or worn brake drum
	Dirt or any other foreign material embedded in the lining. Metallic sound

Note: Breather tube should be kept clean without any blockage. If blocked then friction disc will get soaked in oil.

3.5

Wheels

3.5.1. Functions

The two primary functions of tyres, especially rear ones, are to provide maximum work output while ploughing the field and for haulage.

3.5.2. Size

Front wheel	Rear wheel
6.0 x16,6 PR	12.4 x 28, 8 PR
6.5 x 20, 6 PR	13.6 x 28,12 PR
7.5 x 16,6 PR	14.9 x 28,12 PR
	16.9 x 28,12 PR
12.4 x 24 (4 WD)	18.4 x 30,8 PR

3.5.3. Inflation

Under

If a tyre has insufficient pressure to support the casing, this may be deflected to such an extent that the plies may become separated. Such damage is irreparable and the tyre must be replaced. Excessive deflection allied to a high drawbar pull can cause wrinkling of the tyre side walls which if allowed occurring continually can cause the tyre to creep on the rim and tear out the valve. A visible warning of under-inflation is uneven wear of the lug bars, indicated by 'gouging' of the centre of the bars.

Over

Excessive tyre pressure deprives the tyre of its self-cleaning properties. This causes wheel spin, which in turn causes sinkage, thus increasing rolling resistance of the tyre and lowering the power available for traction. Frequent and prolonged bouts of wheel spin cause wear of tyres. Another effect of over-inflation is that the thread and sidewalls of the tyre are very susceptible to damage from sharp rocks or similar objects due to inability of the casing to 'give' on contact.

3.5.4. Maintenance

Do's for tractor tyres

- Clean the tyre rim with a wire brush to remove the rust
- Always use a new tube with a new tyre
- During fitment, ensure that the tyre rotates in the direction of the arrow on sidewall.
- Check the inflation once a week when the tyre is cold
- Always use dust caps to prevent entry of dust and mud
- Inspect tyres periodically for cuts to be repaired on time
- Keep tyre free from oil and grease
- When tyres are not in use for a long period, jack up the tractor
- Keep tractor in a mechanically fit condition, especially wheel bearings, steering mechanism and brakes
- To avoid slippage during field work, either add wheel weights or hydro inflate the tyres
- To minimise stubble damage, adjust track width

Don'ts for tractor tyres

- Do not check the tyre pressure when the tyres are hot
- Never overinflate or under-inflate the tyre. It will reduce the life of the tyre
- Do not neglect cuts in the tyre
- Do not overload the tyre

3.5.5. Ballasting

Purpose

This is done to provide extra weight on the rear wheels of the tractor for better traction and minimize wheel slippage.

Liquid

- Refers to increasing weight of rear tyres with good, clean water where there is no possibility of freezing.
- To prevent tyre damage by frost, calcium chloride should be dissolved in water to fill the tyres thus forming an anti-freeze solution
- The calcium chloride used should be of commercial grade 70–72%
- Mixing calcium chloride water solution
- To prevent acidity 2.2 kg of lime must be added to every 220 kg of calcium chloride
- Never pour water on to calcium chloride. Always add calcium chloride to water
- Never attempt to add pure calcium chloride to a tyre filled with water as the resultant heat and expansion can cause tyre damage
- When calcium chloride and water are mixed, a chemical reaction causes great quantities of heat to be produced. The resultant solution should be naturally cooled before use.

75% ballasting

- Raise the rear wheels just clear off the ground, using a workshop jack capable of lifting 3 tons or a tractor jack
- Ensure that the tyre valve is secured to the rim, either by a mounting cone or valve nut. After this, deflate the tyre
- Check that an air water type of valve core is fitted to the valve
- Position the valve by turning the wheel at the 12 o' clock position (i.e. vertical and at the top)
- Connect the water adapter to the valve and place the solution suction tube in the tank of solution

- Pump the solution into the tyre until a steady stream pours from the breather hole. This indicates that the tyre has been filled up to the level of the valve, which is approximately 75% of the tyre's capacity.
- Disconnect the water adapter
- Using a special air water gauge, adjust the air pressure to that recommended for the load being carried by the rear of the tractor
- Water ballasting – 75% fill

Tyre size	Additional weight	Capacity of solution (litre/gallon)
12.4 x 28	137 Kg (303 lbs)	127 litres (28 gal)
13.6 x 28	177 Kg (390 lbs)	162 litres (36 gal)

100% ballasting

- A motorized pump is essential for 100% ballasting
- Raise the rear of the tractor using either a workshop jack or a tractor jack until the wheels are just clear off the ground
- Ensure that the valve is secured to the rim by either a mounting cone or valve, Afterwards, deflate the tyre.
- Ensure that an air water type of valve is fitted to the valve body
- Position the valve by turning the wheel until the valve is at the 6 o' clock position (i.e. vertical and at the bottom)
- Fit an adapter extension tube to the special type of water adapter. The length of the extension tube depends on the size of the tyre being filled
- Fit the adapter and extension tube to the valve
- Pump the solution into the tyre until it pours from the breather hole
- Turn the wheel until the valve is in the 12 o' clock position
- Continue pumping until the solution again pours from the breather hole. When this occurs the tyre should be 90-97% full
- Remove the tyre and extension tube

3.5.6. Key factors for ballasting

The most important factors to be considered for ballasting the tyres are the horsepower and the speed of operation.

Generally, higher travel speeds require less ballast, and the mechanical properties of soil only allow so much deformation (i.e. slip) to occur in a given period of time. For example, a faster rolling tyre has less time to cause deformation on a given area of soil; slip is reduced and less ballast is required.

Most tractors are good for 6.5 to 7 km/hr at full power and optimum slip. At 8 km/hr, a very rough starting point for estimating the ideal total tractor weight (including ballast) is around 79 kg/kW. But with change in speed for various operations, the weight varies to maintain the same slip percentage. For example, 6.5 km/hr requires 99 kg/kW for proper ballast; at 9.5 km/hr, only 66 kg/kW is required.

3.5.7. Wheel slippage

The maximum allowed wheel slippage percentage for a two wheel drive (2WD) is 10 to 15% and for a four-wheel drive (4WD) 8 to 12%. These measurements apply only when the tractor delivers full power.

Over ballasting is by far the most common error. If the tractor is over ballasted, it burns more fuel than it should and will experience premature drive train problems.

An under-ballasted tractor is also really no better. This condition wears tyre tread at an accelerated rate due to excessive slip while never really delivering full horsepower to the drawbar. Fuel is wasted because of the extra wheel revolutions made to cover the same distance.

The 8 to 15% guideline provides efficient fuel consumption, tyre life, productivity and component durability.

3.5.7.1. Calculating wheel slippage

- Put a single dot on the side of a rear wheel with spray paint, chalk or tape so that it is easily seen from the side of the tractor.
- Operate the tractor in your usual gear and throttle the setting for the implement. With the implement in the ground, allow enough space to get the tractor up to speed.
- Mark the starting point of the tractor when the chalk mark is at the predetermined position.
- Count ten revolutions of the tyre with the marked point reaching the same position at each revolution. Place the finish stake at this point.
- Raise or detach the implement, return to the starting position and get ready to drive the course again in the same gear, but not in the previously worked ground.

- The position of the tyre mark should be noted when the tractor passes the starting point.
- Again, count the number of wheel revolutions needed to cover the distance between stakes.
- Estimate the wheel revolutions to the nearest 1/4 turn.
- Calculate the percent slip using the formula:

% slip = (10 revolutions with load) - (revolutions without load) x 100/(10 revolutions with load)

- 8 to 8 ½ revolutions without load will give you the optimum 10 to 15% wheel slip. Add or remove weight to obtain the proper wheel slip.

3.5.8. Trouble shooting (Rear wheels)

These are the troubles faced by tractor owners of any make pertaining to rear wheels

Fault	Probable causes
Wheel spin	Wrong gear selection
Wheel spin due to tyres loaded with soil	Excessive tyre pressure. Inadequate tyre pressure. Insufficient weight acting on the rear end of the tractor. Inadequate weight on the front end of the tractor.
Wheel spin (The tyre retains its self-cleaning action and sinks into the ground)	Too narrow section tyre for the weight being carried by the rear end of the tractor. Lug-bar type of tyres being used in the sand.
The tractor sways from side to side when being driven on hard ground (e.g. road)	Excessively low pressure may result in rapid tyre wall wear and consequent failure.
Tyre thread worn unevenly when used for long periods on the road	Too low pressure. Overloading.
Uneven tread wear	Over inflation. Wheels running out of circumference or face.
Tyre creep	Too low tyre pressure.
Split side wall	Under-inflated tyres striking a sharp object.

3.6

Steering Mechanism

3.6.1. Function

It steers the tractor in the required direction. The rotation of the steering wheel moves the steering column, which in turn rotates the sector shaft. This rotation pulls the pit man arms on one side and pushes at the other side. This motion is conveyed to the wheels by the drag link and makes the front wheel turn in the requisite direction.

3.6.2. Components

- Steering wheel
- Steering shaft
- Steering gear
- Pit man arm
- Drag link
- Steering arm
- Tie rod
- King pin

3.6.3. How does it work?

When the operator turns the steering wheel, the motion is transmitted through the steering shaft to the angular motion of the pit man arm through a set of gears. The angular movement of the pit man arm is further transmitted to the steering arm through the drag link and tie rods. The steering arms are keyed to the respective kingpins which are an integral part of the stub axle on which wheels are mounted. The movement of the steering arm affects the angular movement of the front wheel. In other designs, instead of one pit man arm and drag link, two pit man arms and drag links are used and the use of tie rod is avoided to connect both steering arms.

3.6.4. Types of steering gear box

The various types of gears normally available on different makes of tractors are as follows.

a. Worm and nut type

The steering shaft has square threads and is supported in self-aligned bearing; the looseness of the shaft is adjusted by tightening or loosening the cushioned rings on the steering outer shaft. A nut having square identical threads is placed on the worm. This nut is hinged to the sector shaft. On moving, the steering shaft nut moves up and down making the sector shaft and drop arm move to and fro.

b. Worm and sector type

Worms are cut on the shaft. This shaft is supported in the housing with the help of two bearings one at the top and the other at the bottom and can rotate easily. A sector with identical teeth is held in the sector shaft. The sector shaft rotates the two bronze bushes on either side of the steering housing. On moving the steering wheel, the worm shaft along with the sector shaft moves resulting in the movement of the drop arm.

c. Worm and roller type

It is an improved version of the worm and nut type of steering. In this type, steel balls are placed in the nut having semicircular threads. As such, the balls run half in the threads cut in the worm, while the other half runs in the recess cut in the nut. The balls are fed from one end and return back through a pipe clamped on the nut. The set of balls go on circulating and minimize friction. The nut in which the balls move is hinged on to the sector shaft which runs on to bushes fixed in the steering housing. When the worm moves, it takes the nut and the sector shaft along with it and the drop arm moves to and fro.

3.6.5. Power steering

a. Components

- Pump
- Cylinder mounting plate

- Steering cylinder
- Fluid tank
- Steering arm link
- Supply hose
- Pressure hose
- Return hose

b. How does it work?

Power steering is actually "power-assisted steering". A pump driven by the engine delivers pressurized hydraulic fluid to either side of the steering mechanism. This pressurized fluid acts on the steering mechanism and helps to steer the tractor and reduce the steering effort. The pressurized fluid is fed into the double acting steering cylinder (independent for each wheel). The flow to the cylinder is controlled by the steering wheel with the help of valves. Power-assisted steering is not a drive-by-wire system. It just helps the driver to steer the vehicle with minimum effort.

There are two main parts to the power steering system: the pump and the steering gear. In most cases, the pump is attached at the front end of the engine and is driven by the accessory drive belt/gear train. The fluid reservoir is usually located on the pump. The pump should be sized to deliver sufficient fluid pressure at idle. As the pump spins faster as the engine speed increases, a pressure relief valve is used to keep the pressure at the desired levels. In some cases, the engine driven pump may be replaced by an electric pump. The major type of steering gear used in tractor is the worm and roller type.

3.6.6. Hydrostatic steering

a. Components

- Steering control unit
- Flow amplifier valve

b. How it works?

When the steering unit is activated, a controlled oil flow is directed to the flow amplifier valve. This oil flow is amplified and the total flow is directed to the steering cylinders. The steering unit provides a fixed displacement of oil per revolution of the steering wheel and the amplification factor of the flow divider valve is 8. Hence, the total oil output is eight times the output of the steering control unit.

With this system, it is possible to combine the steering and working of hydraulics. A priority valve is built into the flow amplifier valve which ensures that the steering has first priority on oil flow from the hydraulic pump. If the oil flow is not used for steering, the entire oil is sent via the "EF" line (external flow) to the working hydraulics at minimal pressure loss.

The principle applied to the controlled operation of this system is called load sensing. As the name suggests, it is a system in which the load is sensed or registered. The sensed signal is used to control the priority valve in the flow amplifier valve, so that oil flow and oil pressure precisely match momentary demands. Technically, what happens is that the load signal (a pressure) is registered between the load (steering cylinder) and the downstream side of a variable metering orifice in the steering unit. This signal (pressure) is led to the priority valve spring chamber which regulates the oil flow so that the pressure drop across the metering orifice is always constant. A schematic of working of hydrostatic steering is presented (Fig 3.6.1).

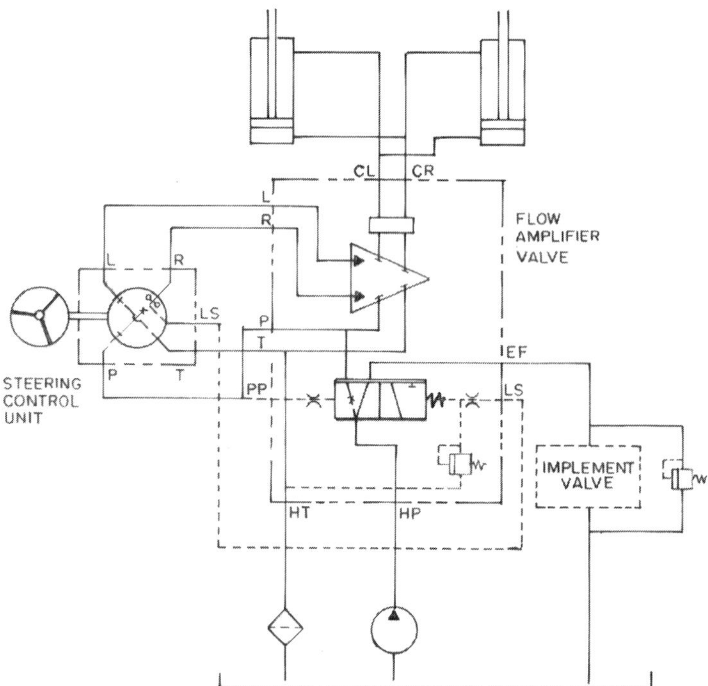

Fig 3.6.1. Hydrostatic steering

P- Pump, T- Tank, L- Left, R-Right, CL- Cylinder left, CR-Cylinder right, EF-Excess flow, LS-Load sensing signal, PP- Pilot pressure, HP- High pressure from pump, HT-high pressure to tank

3.6.7. Trouble shooting

These are the troubles faced by tractor owners of any make pertaining to steering

Fault	Probable causes
Steering vibration	Check for wheel alignment. Steering slackness. Wear of tie rod end. Spindle play. Loose joint linkages. Spindle bearing, hub and bearing condition. Wheels bend. Bend at the centre of front axle
Steering shock	Incorrect tyre pressure. Inner column nut loose. Worn out primary and/or secondary shaft
Hard steering	Incorrect tyre pressure. Incorrect lubrication. Worn out linkages. Toe in adjustments. Steering components
Free play of steering	Gear box not properly fitted. Worn out linkages. Worn tie rod ends. Backlash on primary/secondary shaft.

3.7

Front Axle

3.7.1. Function

The front axle is the unit on which the front wheel is mounted. This wheel is an idler wheel by which the tractor is steered in various directions. The function is to assist the steering and support the tractor. It is a rigid tubular or I-section steel construction, pivoted at the centre. The pivot pin is housed on bushings on a support bolted to the front of the engine. As the track width of a tractor is adjustable, the front axles have provision to be extended on both sides. The tubular front axles have telescopic arrangement for extension, whereas the I-section type of front axles are three-piece axles.

3.7.2. Types

Swept and straight

3.7.3. Main components

- The front axle assembly consists of a centre beam and two outer axles
- The centre beam pivots on a pin which is bolted to the front engine support
- The front axle (RH/LH) can be bolted to the centre beam in alternative positions to provide front wheel track adjustments

3.7.4. Front axle adjustment

Track width adjustment

Depending on the requirement of the tract width, the axles can be extended on both sides. Whenever the track width is adjusted, the wheel toe-in must also be checked and corrected.

Toe-in adjustment

The front wheels are slightly drawn in at the front side in such a way that it is slightly less than the rear side. This difference is known as toe-in and varies in the range of 4±2mm.

To get the desired toe-in, the procedure followed is as follows:

1. Bring the tractor on a levelled ground with front wheels in the straight-ahead position. This can be done by moving the steering wheel from one extreme to the other extreme position and then bringing it to the middle of its total revolution.
2. Mark the centre point on the width of each tyre and measure the distance between both the front types at the front and rear sides. For desired results, the rods or drag-link ends are slackened and the rods are rotated clockwise or anti-clockwise to increase or decrease the toe-in. It must be ensured that the distance of both the tyres is equal from the centre of the axle. Similarly, the rear side of both tyres should be equal from the centre of the axle, but more than the distance at the front side. To achieve equal movement, both the drag links should be rotated equally.

Caster angle

It is the angle between the centre line of the king pin of the tractor and the vertical line.

Camber angle

It is the angle between the vertical axis of the wheel and the vertical axis of the vehicle when viewed from the front or rear. It is used in the design of steering and suspension. If the top of the wheel is farther out than the bottom (that is away from the axle), it is called a positive camber; if the bottom of the wheel is farther out than the top, it is called a negative camber. To get the maximum straight line acceleration, the greatest traction will be attained when the camber angle is zero.

3.7.5. Trouble shooting

These are the troubles faced by tractor owners of any make pertaining to the front axle

Fault	Probable causes
Wobble in front wheel	Worn out king pin bushes and taper roller bearings of the wheels
Tractor drag, steering vibration	Bent front axle

3.8

Power Take-Off (PTO)

3.8.1. Function

A power take-off (PTO) is a splined drive shaft on a tractor that can be used to provide power to an attachment or a stand alone machine. PTO shaft gets the drive from hydraulic pump shaft. This in turn gets drive from engine. PTO is designed to be easily connected and disconnected (Fig 3.8.1.). It is used to operate implements, *viz.,* rotavator, fertilizer spreader, duster and many more machines which are stationary such as threshers, water pump and portable flour mill. These shafts either run at 540 rpm with 6 splines in the shaft or at 1000 rpm with 21 splines in the shaft. But many old tractors have 540 rpm PTO. The present day equipment is designed to operate with 1000 rpm PTO shaft. The manufacturers of old tractor models provide an auxiliary gear box which increases the speed of PTO to 1000 rpm. Selection of implements, equipment and machinery should be done on the basis of available tractor PTO speed.

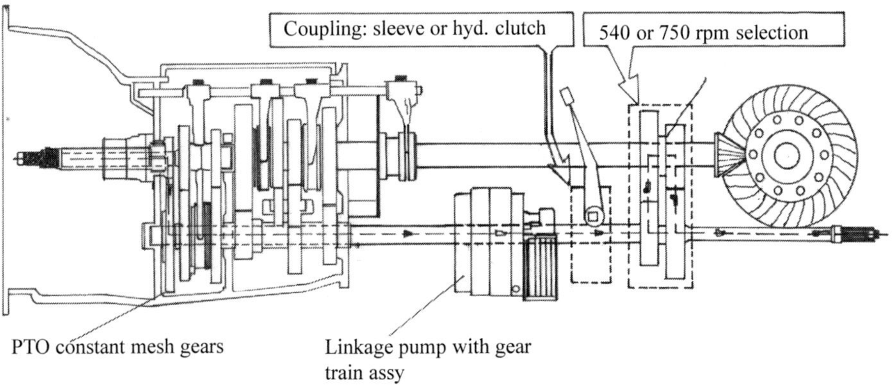

Fig 3.8.1. Power Take-Off

3.8.2. Main components

- Hub PTO drive
- Bushing PTO
- Bearing needle roller
- Shaft assembly
- Circlip
- Ball bearing
- Seal Special - PTO shaft
- Housing -PTO seal
- 'O' Ring
- Cap PTO
- Fork assembly
- Cover PTO shift lever
- Assembly lever- PTO

3.8.3. Types

a. Two-speed PTO

Provides two different PTO speeds for a given set of ERPM (Engine Revolution per Minute) for various PTO operations. In general, 540, 740 and 1000 are the preferred/optimized PTO speeds which cover most of the applications and meet the required speed of operation.

ERPM	PTO RPM	Application
1500	540	Ideal for general operations
2000	540	Rotary tiller
2000	750	Sprayers
2000	1000	Stationary applications such as threshers, generators, pumps, electric/ saw mill wood cutting application

b. Reversible PTO (Fig 3.8.2.)

Suitable for operating PTO-driven implements: harvester, straw reaper, baler, etc. Whenever cut plant parts like straw hinders the rotation of the blades, the tractor engine struggles to continue to run the implement. Under such conditions, without stopping the equipment by operating the reversible PTO, the blades are made to run in the opposite direction, thereby releasing the struck plant part from the blade without the driver necessarily coming down.

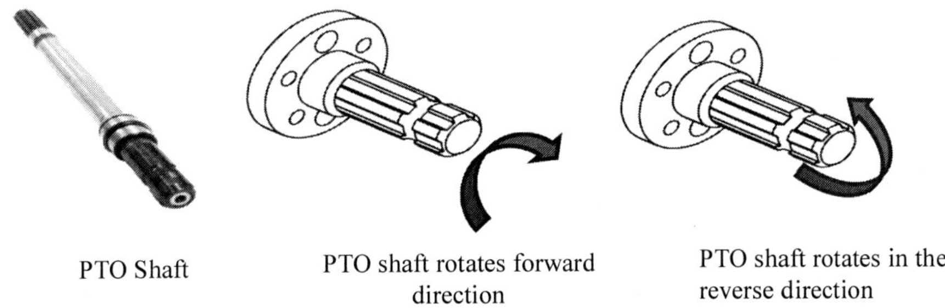

PTO Shaft | PTO shaft rotates forward direction | PTO shaft rotates in the reverse direction

Fig 3.8.2. Reversible PTO

3.8.4. Trouble shooting

These are the troubles faced by tractor owners of any make pertaining to PTO:

Fault	Probable causes
Wear and tear of "O" ring and seal in PTO assembly	Usage over a period of time

4

Hydraulics

4.1. Hydraulic system

It is a mechanism in a tractor to raise, hold or lower the mounted or semi-mounted equipment by hydraulic means. All the present day tractors are equipped with a hydraulic control system for operating three point hitches. Some of the tractors are fitted with a hydraulic brake system and a few others are fitted with a hydraulic steering system. Tipping trailers, front mounted loaders or dozers are examples of multi use of hydraulics.

4.2. Principle

The working principle of a hydraulic system is based on Pascal's law (Fig 4.1.) This law states that the pressure applied to an enclosed fluid is transmitted equally in all directions. A small force acting on a small area can produce higher force on a surface of larger area.

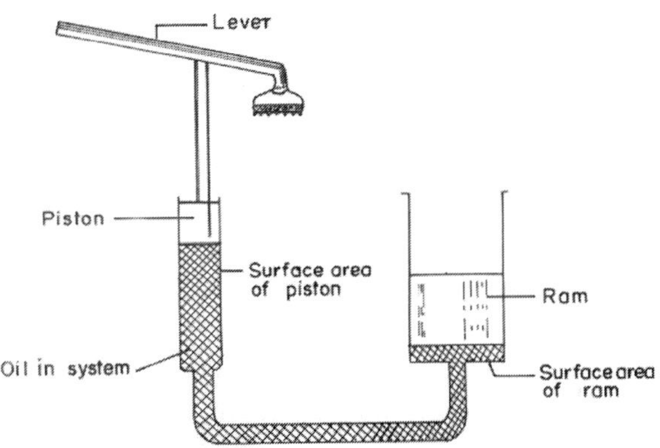

Fig 4.1. Working of hydraulic system

4.3. Components and working

A simple hydraulic system consists of the following basic components, *viz.*, reservoir, pump, relief valve, control valve and cylinder.

4.3.1. Reservoir

It contains sufficient oil to move the piston for lifting the load.

4.3.2. Pump

Pumps displace fluid, causing a flow. Adding resistance to flow causes pressure. The hydraulic pump in some tractors is located at the centre housing and submerged in oil. It supplies oil under pressure to the system for the operation of implements and equipment. The pump is supported as follows:

(a) The sides are supported by pump pin supports from both sides of the centre housing.

(b) The front portion (camshaft) is supported and connected to the countershaft of the gear box by a coupler.

(c) The rear portion (camshaft) is supported on the PTO shaft with the help of a bearing.

The pump is driven by the countershaft and it in turn drives the PTO, if coupled to-gather. The speed of the pump varies with the speed of engine. The pressurization of oil takes place when the control lever is operated

4.3.3. Relief valve

It protects the system from high pressure. The valve is set for slightly higher than the working pressure. In case the pressure increases beyond the working pressure, the relief valve opens allowing the fluid to pass on to the reservoir.

4.3.4. Control valve

The control valve is used to allow the operator to direct the flow of fluid either from the pump to the ram cylinder or from the cylinder to the reservoir. It has four positions, namely lift, neutral, slow drop and fast drop.

4.3.5. Ram cylinder

The ram cylinder converts the hydraulic power to mechanical power for doing the various jobs.

4.4. Different pumps used in tractors

4.4.1. Gear type

These pumps are fixed displacement types, meaning they pump a constant amount of fluid for each revolution. They are one of the most common types used for hydraulic power application. These pumps create pressure through the meshing of the gear teeth, which forces fluid around the gears to pressurize the outlet side. These are fitted outside and derive power.

4.4.2. Axial piston type

An axial piston pump has a number of pistons (usually an odd number) arranged in a circular array within a housing which is commonly referred to as a cylinder block, rotor or barrel. This cylinder block is driven to rotate about its axis of symmetry by an integral shaft that is, more or less, aligned with the pumping pistons.

4.4.3. Radial piston type

This type includes fixed displacement pumps that are used for high pressure and relatively small flows. Pressures of up to 650 bars are normal. Sometimes, these pumps are designed in such a way that the plungers can be switched off one by one, so that a sort of variable displacement pump is obtained.

They produce a very smooth flow. In variable models, the flow rate changes when the shaft holding the rotating pistons is moved with relation to the casing (in different models either the shaft or the casing moves). The output can also be varied by changing the rotation speed.

4.4.4. Auxiliary pump

This delivers pressurized hydraulic fluid to operate auxiliary equipment or attachments. The addition of an auxiliary hydraulic system to heavy construction equipment increases the versatility of the vehicle by allowing it to perform additional functions with different attachments.

The auxiliary hydraulic system may vary depending upon the vehicle, e.g. an excavator, a back-hoe or a front-end-loader. Vehicle interface fittings, length from the pump to the attachment and vehicle control systems require various configurations of an auxiliary hydraulic system. Auxiliary hydraulic systems usually include external fluid fittings to facilitate connecting and disconnecting the hydraulic fluid supply lines to the vehicles' hydraulic pump. They also usually include valves configured to control the supply of hydraulic fluid through the fittings. Different attachments also have varying requirements of an auxiliary hydraulic system with respect to timing, rate and control of fluid flow.

4.5. Working of hydraulic

The implement is linked directly to the tractor so as to make it one integrated unit. The implement is controlled hydraulically. The linkage system is referred to as "three-point linkage". Three-point linkages involve two links at the bottom and one link at the top. The two bottom links pull the implement, while the top link serves as a rigid brace between the tractor and implement, making it difficult for the front end of the tractor to lift high off the ground. When the implement is pulled back by the bottom links, when on the ground, the top link is pushed forward which in turn activates the hydraulic mechanism. This tends to exert a lifting force on the implement. This lifting tendency adds weight to the rear wheels and increases traction. It also acts as a counter weight to the front end and helps in shifting the front weight to the rear wheels to get traction. The hydraulic mechanism built in the tractor is operated both manually and automatically to adjust the depth of working of various implements and also raise and lower any implement or equipment.

The position control lever settings are used for lifting and lowering. Besides, adjustment is also done suitably during transport of implement. The draft control settings facilitates to adjust working depth and for transfer of weight of the implement and soil resistance to the driving wheels which helps in reducing slippage.

The hydraulic mechanism is utilized to avoid or reduce the danger to the implement and driver that may arise from the implement hitting a solid obstruction. An overload release action automatically reduces the traction or weight on the rear wheels and increases it on the front wheels when such an object is hit. Instead of an extra weight on the rear wheels causing the front end of the tractor to lift, there is reduced traction causing the rear wheels to spin.

4.5.1. Attachment and detachment of implements

Attachment

The tractor is reversed and the hitch pin of the implement is aligned up to the ball ends of the lower link. The position control lever is used to raise or lower the lower links until the left hand ball end aligns with the hitch pin of the implement. The ball is pushed through the hitch pin and secured with the lynch pin. The right hand lower link is attached to the implement. The levelling lever is used to adjust the height of the link if necessary. One end of the top link is attached to the frame of the implement and the other end to the clevis assembly.

Detachment

A levelled area is selected since it will make it easier for detaching an implement. The implement is then lowered. The tractor end of the top link is then disconnected. The tractor is dismounted and the lower links are detached. The lynch pins are replaced in their position.

4.6. Trouble shooting

These are the troubles faced by tractor owners of any make pertaining to hydraulics

Fault	Probable causes
Implement does not raise	Lift cover installed in such a way that the ends of the vertical lever are not positioned correctly to operate the pump lever. Sticking control valve. Leak in the system. Broken, bent or damaged control linkage. Faulty relief valve. Broken or damaged internal pump parts. Seized ram cylinder. Lift arms binding. Implement weight too great for the system.
Implement lifts, but will not lower	Sticking control valves. Damaged control valve springs. Lift arms binding.
Jerky or uneven lift when operational lever is raised	One or more in-operative side chamber valves.
Relief valve blows when operational lever is raised to transport position	Quadrant stop not properly positioned. Check chains are twisted. Check chains installed in the lower holes of anchor brackets. Lower links reversed.
Implement will not lower or will raise when the operational lever is moved to the lower stop	Stop incorrectly located. Valve lever adjusting nut is too tight. Self-locking adjusting nut on the horizontal spring guide for the position control linkage too loose.
Unable to obtain slow drop when the operational lever is moved into the response sector of the quadrant	Eccentric cam on vertical position control lever not properly positioned with respect to the position control cam arm. Self-locking nut on the horizontal spring guide adjusted too tight after adjusting the eccentric cam.

5

Instruments and Controls

5.1. Instruments panel

5.1.1. Oil pressure gauge

Indicates oil pressure in Kgf/cm^2

Colour range	Pressure
Green	Normal working pressure
Red	Low oil pressure

5.1.2. Hour meter cum RPM meter

- Indicates engine speed in rpm and helps to carry out maintenance schedule
- Registers one unit for every hour of work that the engine performs at 1500 rpm
- The rectangular aperture in the dial shows the reading of an odometer which registers hours of engine worked

5.1.3. Water temperature gauge

Indicates engine coolant temperature

Temperature range (^{0}C)	Indication
80-105	Optimum working temperature range
Above 105	Red range. Over heating

5.1.4. Throttle lever

The lever is moved down to increase the engine rpm

5.1.5. Horn button

Button is pressed to sound the horn

5.1.6. Heater starter switch

It has 4 positions

Temperature range (^{0}C)	Indication
O (Off)	-
ON or ACC (Auxiliary start)	Turning the key to the start position, supplies power to the starter motor
H (Heat)	Turning the key to this position supplies power to the auxiliary lighting units.
HS (Heat/Start)	Turning the key to the heat/start position keeps the thermostat coil energized while supplying power to the starter motor

5.1.7. Lighting switch

The lighting switch has 5 positions. Operated by turning the knob in a clockwise direction

Position	Indication
O	Off
1	Side, Tail and Panel
2	Side, Tail, Panel and Dipper
3	Side, Tail, Panel and Main Beam
4	Main Beam only

Plough lights and horn operate in all the 4 positions

5.1.8. Fuel gauge

Mark	Indication
Green band	Fuel level safe zone
Red band	Fuel level in tank nearing empty
E	Fuel tank empty
½	Fuel tank filled up to half level
F	Fuel tank completely filled with fuel

5.1.9. Fuel cutoff control

- To stop the engine, the knob is pulled fully outwards
- The knob is pushed fully in before starting the engine

5.1.10. Indicator switch

Indicator switch has 3 positions

Position	Indication
Centre	Off
Left	Left side turn indicator lights operates
Right	Right side indicator operates

5.1.11. Ammeter

Indicates rate of charge/discharge of battery.

- If needle is 0 then it indicates battery is fully charged
- While starting the tractor, the needle reflects 0-30 (discharging red)
- While the engine works the needle reflects 0-30 (charging green)
- While engine is working and needle moves and light on the needle moves the left side (-). It is an indication that the battery is not getting charged.

5.1.12. Hazard warning light switch

When the switch is operated, all the four indicator lights will start flashing simultaneously. This is to warn other road users of the potential danger.

5.1.13. Fuse box

The fuse in the fuse box protects various electrical circuits.

5.1.14. Dual range selector lever

- Used to select high or low transmission range
- On the knob, the L–S–H mark indicates LOW, START (Neutral) and HIGH, respectively
- The lever must be in the 'S' position before the engine is started

5.1.15. Gear shift lever

- Used for selecting the required gear.
- Enables selection of the required gear
- Marking on the gear shift lever knob indicates the respective shift position

5.2. Seat adjustments

- Provided with a deluxe seat
- To suit the operator, height and weight adjustments are provided
- The slot in the mounting bracket allows the seat to be moved forward or backward
- The slot in the back rest bracket allows the height of the back rest to be increased or decreased
- The adjusting knob at the back of the seat stiffens or softens the spring suspension when turned clockwise or anticlockwise, respectively

5.3. Left hand side controls

5.3.1. Clutch pedal

- The clutch pedal has to be depressed to disengage the transmission from the engine
- While the tractor is started, the clutch pedal is depressed

5.3.2. Parking brake lever

- Used along with a foot brake to park the vehicle safely, so that it does not move.
- Acts on tractor rear wheels
- To engage the brake, the foot brake pedals are pressed down and the hand lever is pressed up
- To release the parking brake, the foot brake pedal is pressed down, followed by pressing the button at the crown of the lever to push the lever down.

5.3.3. PTO lever

- The PTO lever when pulled rearward engages the PTO shaft and the shaft rotates at a speed of 18/50 of the engine RPM. When in neutral, the shaft is disengaged.

5.3.4. Fuel tap

- Usually located on the left hand side of the tractor beneath the fuel tank
- The tap is turned clockwise to cut off fuel supply

5.4. Right hand side controls

5.4.1. Brake pedals

The two brake pedals can be used independently or latched together to prevent the master pedal from normal braking. For independent brakes, the latch is disengaged. The inner pedal acts on the left hand rear wheel. The outer pedal acts on the right hand rear wheel.

5.4.2. Foot throttle

- Operation of the foot throttle overrides the hand throttle setting when increasing the engine speed
- When the foot throttle is released, the engine will return to the speed set by the hand throttle that must be fully closed.

5.5. Hydraulic quadrant controls

The Ferguson hydraulic system combines the tractor and implement into one unit with the implement hydraulically controlled. The system fulfils the following functions:

- Depth control of soil engaging implements (Draft control)
- Control and positioning of implements above the ground (Position control)
- Control of external equipment having hydraulic operation—trailer tipping, loader operation, hydraulic motor drive, etc. The speed of the implement drop can be controlled by the response control
- The hydraulic control lever quadrant is located on the right hand side of the tractor seat within easy reach of the operator.

5.6. Quadrant controls

A tractor hydraulic lift control system is provided with a manually adjusted quadrant lever for determining the draft at which the tractor is to operate and a second manually adjusted quadrant lever for determining the implement height. The draft control may raise the implement above the height determined by the position control, but may not lower it below the height determined by position control.

5.6.1. Position control

Position control lever (Operational lever) in its upper range of quadrant raises and lowers the implement besides providing position of the lower link height. The operator can select and maintain a fixed height or depth of an implement. An adjustable stop is provided on the quadrant so that the implement returns to its previously selected position.

5.6.2. Response selection

To suit the implement, its rate of drop is made variable by the position of the operation lever in its lowest range which is suitably marked as "FAST–RESPONSE–SLOW".

In case of heavy soil or when there is bouncing of implement, then the operational lever is moved closer to slow the response position. The final setting should be locked with the knob as the operational lever is used to raise the implement at the end of the furrow. It should also be ensured that the implement does not pitch with the tractor.

5.6.3. Draft control

An implement is lowered into the working position by moving the operational lever down through the position control range. The draft control lever is used to select the desired depth. The implement will penetrate, when a low level is set down the quadrant. Setting should begin with the lever in line with 2 dots stamped on the quadrant plate. The small adjustable selector is positioned in line with the lever setting and locked using the knob. The small sector defines its working range on either side to overcome slight changes in the depth of penetration of implement.

5.6.4. Lever settings of the external hydraulic equipment

Quadrant control level setting for various operations

a) Transport position

The operational lever is raised to the top of its quadrant. When transporting implement for a distance, the draft control lever should be at the bottom of its quadrant. To lift the implement at the end of the furrow and while ploughing, the draft control lever - I is left at the chosen setting.

b) Normal ploughing and cultivating

The operational lever is kept in the response sector according to the response required. The draft control lever is kept below the sector marks at the bottom half of its quadrant according to the required depth.

c) Shallow cultivation - heavy implement

The operational lever is kept in the response sector according to the response required. The draft control lever is kept below the sector marks in the bottom half of its quadrant according to the required depth.

d) External hydraulic equipment

The operational lever is kept in the response sector as required. The draft control lever is to be kept above the sector marks. This will provide oil flow for operating the external equipment. The movement of lever below the 'sector marks' will allow return of oil from the external equipment.

e) Position control-soil engaging implements

The operational lever is kept in position control range according to the soil on which the implement is required to operate. The draft control lever is kept at the bottom of its quadrant in the maximum draft position.

f) Position control- above ground implements

The operational lever is kept in position control range according to the height at which the implement is required to operate. The draft control lever is kept at or anywhere below the sector marks.

5.7. Summary of lever settings (example drawn from Mark 3 hydraulic-MF tractor)

Particulars	Position control lever	Draft control lever	Response control lever
Transport position	Place at the top of the quadrant in the 'UP' position	Place at the top of the quadrant against stop in the 'Transport' position	Not used
Ploughing and cultivating (Entering work)	Move the lever to the 'DOWN' position to lower the implement for work. As the tractor moves forward, the implement will penetrate. Move the lever further to the 'Down' position to make the implement go deeper into the soil.	Not used. Place the lever back against the 'Transport' stop and lock it there with knob to prevent accidental movement. Place the arrow indicator on the SLOW side of the centre position of the response quadrant	
Ploughing and cultivating (Working)	When the desired working depth is obtained, move the adjustable stop level with the draft lever and lock into position with a knurled nut. The draft lever can be moved slightly on either side of the knob to suit varying soil conditions	Not used. In transport position	If the control depth varies (over ridge and furrow, for example) move the arrow indicator towards the 'FAST' position
Ploughing and cultivating (Leaving work)	On reaching the head space of the field, raise the implement by moving the draft control lever back to the 'UP' position. Re-enter field by moving the draft control lever forward to the adjustable stop	Not used. In transport position	Lever as in earlier position
Position control (Entering work)	Not used. Place lever at the top of the quadrant in the 'UP' position	Move the lever forward in the position sector until the linkage carries the implement at the height for work	Not used. In the middle of range

Position control (working)	Not used. In the 'UP' position	When the desired working height is obtained, move the adjustable stop up to the position control lever and lock in position with the knob	Not used. In the middle of range
Position control (leaving work)	Not used. In the 'UP' position	On reaching the headspace, raise the implement out of work by moving the position lever back to the 'Transport' stop. Re-enter work by moving the position control lever forward to adjustable stop.	Not used. In the middle of range
External hydraulic power operation (when spool valves are not fitted-suited for loaders and tipping trailers)	Move the lever back past the 'Transport' stop into the 'Constant pumping' sector. The lever must be pushed to the side to overcome the stop gate.	Move the lever to a point about three-quarters back on the quadrant until a position is found approximately in the sector marks, which holds the external ram stationary .Then, move the adjustable stops to line up with the lever and lock in place with the knurled nut	
External hydraulic power operation (working)	Raise the loader, for example, by moving the lever back and lower by moving the lever forward. The rams can be held stationary by moving the lever to the adjustable stop	Not used. Lever right back in constant pumping	Control the rate of drop with the response control lever, moving the lever forward towards 'FAST' for a rapid drop.

6

Electricals

6.1. Types

Lamps: Head lamp, plough lamp, indicators

Switches: Safety switch, starter switch, etc

6.2. Components

Battery: A battery provides power to electrical or electronic circuits. The voltage rating of a battery tells how much electrical "pressure" the battery can provide. Battery energy is used to operate starter motor and to provide current to the ignition system during cranking. The electrolyte should always be maintained just below the maximum level, but should not go down below the $1/4^{th}$ level.

Alternator: Alternators are electromechanical devices used to convert mechanical energy to alternating current electrical energy. While some of the alternators use rotating magnetic field, some use linear. The functioning of alternators is similar to that of direct current generators. As the magnetic field around a conductor changes, a current is induced in the conductor. The rotating magnet or the rotor turns and cuts the field across the conductors. Thus, an electrical current is produced which causes the rotor to turn. This rotating magnetic field causes an AC voltage in the stator windings. In tractors, alternators are used to charge the battery and to power the electric system when its engine is running. An alternator is superior to a dynamo as it provides more precise output control, requires lesser maintenance, is more compact and requires no cut-out.

Starter motor: Starting motor or starter is an electric motor that rotates an internal combustion engine to cause the engine to begin powering itself. When the ignition key is turned to the "Start" position, the battery voltage goes through the starter control circuit and activates the starter solenoid, which in turn energizes the starter motor. The starter motor cranks the engine. A starter is usually operated when the clutch pedal is depressed in many tractors.

An advanced starter motor is the gear reduction starter motor. This type of starter motor contains a magnetic switch, a compact high-speed motor, reduction gears, a pinion gear and a starter clutch. The extra gears reduce the motor speed by a factor of one to three or four and transmit it to the pinion gear. The plunger of the magnetic switch directly pushes the pinion gear which is located on the same axis, causing it to mesh with the ring gear in the engine.

Neutral safety switch: This is an important electrical component. This switch disables the starter operation when the vehicle is in the gear position. If the tractor is allowed to start in gear by an inexperienced person, it would immediately begin to move and might result in accidents. Neutral safety switches are equipped to ensure that the vehicle starts only when the shift lever is neutral.

Fuse box: It is located on the battery platform and controls the amount of power that moves from the battery to specific devices in the tractor, *viz*., head lights, hazard warning light, parking light, horn, brake light and plough lamp. Its function is to protect various electrical circuits against surges.

Wiring harness: Also known as cable harness or cable assembly or wiring assembly, it is a string of cables and/or wires which transmits informational signals or operating currents (energy). The cables are bound together by clamps, cable ties, cable lacing, sleeves, electrical tape, conduit, a weave of extruded string or a combination thereof.

In constricting the wires into a non-flexing bundle, usage of space is optimized and the risk of a short is decreased. Harness connects the horn terminals, lamp sockets, temperature sensors, alternator, warning lamp terminal, starter motor, starter solenoid, earth of starter motor, neutral safety switch terminals, etc. Binding the wires into a flame-retardant sleeve also lowers the risk of electrical fires.

6.3. Trouble shooting

These are the troubles faced by tractor owners of any make pertaining to battery, self-starter, dynamo/alternator and temp gauge.

6.3.1. Battery

Fault	Probable causes
Overcharging	High voltage. High temperature
Low battery output	Low electrolyte level. Defective battery cell. Low battery capacity.
Battery uses excessive water	Overcharging. Case cracked. Leakage at cover seal.
Bulged case	Overcharging
Corroded battery	Overcharging. Overfilling
Cracked case	Frozen battery. Battery loose in holder
Low battery charge	Excessive load. High self-discharge. Faulty wiring circuits. Faulty generator or regulator

6.3.2. Self-starter

Fault	Probable causes
Self-starter does not crank engine. Self-starter relay or solenoid does not engage	Battery discharged. Self-starter or solenoid is inoperative. Starting circuit open
Self-starter turns but does not crank engine	Faulty self-starter drive mechanization. Faulty solenoid or pinion engagement levers
Self-starter does not crank engine. Self-starter relay or solenoid will engage	Battery discharged. Defective self-starter or battery connections. Defective self-starter. Solenoid contacts burnt. Engine seizure
Self-starter cranks slowly	Battery discharged. Excessive resistance in starting circuit. Faulty self-starter. Engine rotates too tight

6.3.3. Alternator

Fault	Probable causes
Charge at high rate	Faulty voltage regulator. Defective battery. Faulty dynamo/alternator
No output	Broken drive belt. Defective voltage regulator. Loose connection or broken cable. Faulty dynamo/alternator
Intermittent or low output	Drive belt slipping. Loose connection or broken cable. Defective voltage regulator. Defective dynamo/alternator
Not producing current	Loose drive belt. Defective voltage regulator. Worn out bushes. Short circuit in charging circuit. Defective diodes. Short circuit in rotor
Sound production	Worn out bearing. Loose drive belt. Loose foundation bolts. Loose fan belt. Short diodes
Poor life of regulator	Air gap in regulator points. Loose earth connection. Short in rotor

6.3.4. Temp gauge

Fault	Probable causes
Low temperature (0-40°C)	
If '0' then	Wrong wiring or disconnection in wiring. Defective sensor. Defective gauge
If >'0' <= 40°C	Ambient temperature is low (-15 to 0° C). Defective gauge. Defective thermostat (Always open)
High temperature (>105°C)	Loose/cut fan belt. Defective radiator cap. Fan blade fitted in opposite direction. Less or no water in radiator. Choked radiator (Out side). Blocked tubes in radiator. Defective gauge. Defective thermostat (Not opening).

7

Applications of Farm Tractor

Farm tractor designs and styles differ greatly. Tractors are often used on a daily basis for several agricultural and non-agricultural tasks by attaching or operating light, medium to heavy implements, equipment and machineries according to the requirement and hp of tractors. Tractors are very versatile and can be used on sandy (light textured), loamy (medium textured) and clayey (heavy textured) soil at different topographic and agro-climatic conditions. Earlier, the proportion of tractor usage in agricultural and non-agricultural work was 70:30. This proportion has increased to about 55: 45 over the past two to three years. It is expected to reach 50:50 in (agricultural:non-agricultural applications) in the coming years. The major application of farm tractors for agricultural and non-agricultural sectors is given below.

7.1. Agriculture sector

- Preparing the field by attaching different implements, equipment and machinery for tilling, disking, harrowing, levelling and forming of beds, bunds, furrow and ridges
- Puddling of rice filed by attaching puddling implements, equipment and machinery
- Digging of pits by attaching post-hole digger to plant saplings and erect the poles
- Sowing, planting and transplanting of crops by attaching different types of seed drills, planters and seedling transplanters
- Application of organic manure in line or rows by attaching a trailer and organic manure applicator

- Broadcasting, side dressing and foliar spraying of inorganic fertilizers by attaching the fertilizer broadcaster, ferti drills and sprayers
- Intercultural operation in wide-spaced standing crop by attaching different types of implements
- Spraying of pesticides in field crops and trees by operating different types of sprayers
- Pumping the ground, canal, river, open well, pond or stored water by operating pumps
- Harvesting of crops by attaching different reapers, harvesting machines and tractor power-driven combine harvester
- Pruning of trees by attaching hydraulic-operated pruning tools
- Harvesting of fruits by attaching hydraulic-operated elevators and plucking tools
- Crushing of sugarcane by operating sugarcane crusher
- Threshing of crops by operating threshers
- Winnowing of crops by operating winnowers
- Hauling of water tanker to irrigate perennial trees and nursery
- Hauling of farm inputs and produce
- Transportation of farm workers
- Precision farming by installing the GPS (Global Positioning System) technology in tractor to navigate the farm equipment for sowing, planting, intercultural operation, spraying pesticides and herbicides, applying fertilizers and harvesting accurately
- Mowing and rolling the lawns and golf courses by attaching the mowers and rollers
- Cutting weeds by attaching weed slashers

7.2. Non-agriculture sector

- Developing and levelling of roads, building sites, dams, etc. by attaching dozer blade, bucket, hoe and laser-guided leveler
- Hauling of roads, buildings and construction materials
- Powering of building and road construction equipment

- Using for defense and airport applications to tow baggage
- Removing the snow by attaching a snow blower and grader
- Breaking of rock with Tractor mounted pneumatic compressors
- Generating electricity by operating the alternator/generator

Tractors today accept a wider variety of attachments than ever before, making them increasingly more versatile. Each year, manufacturers offer more new and innovative improvements to their models. There is a still large untapped potential to explore.

8

New Generation Tractors

Features	Details
Instrument cluster (Fig 8.1.)	
Indicators for level of hydraulic and brake oil, selected gear and speed of tractor in instrument panel.	To know the optimum level and right time for refilling or changing of oil.
Revolution (RPM) control	To set the required PTO speed for different PTO-driven implements, equipment and machinery.
Electronic fuel consumption gauge	To record the actual fuel consumed and time taken to utilize the quantity of fuel. This helps to optimize efficient use of fuel for various fields and on road operations.
Virtual terminal	Implement control screen is shown on the tractor virtual terminal when chosen by the operator.Graphics displayed on the virtual terminal are defined by the implement control module.Operator uses soft keys or touch screen to manipulate the implement.
Electronic Control Unit (ECU) (Fig 8.1.)	
Electronic engine control unit	To programme the controls unit (pressure sensor, throttle position sensor and fuel injection) in such a way that it increases the torque of the engine and saves fuel from 15 to 20% . This results in lowering the rpm when working under similar conditions.
Electro-hydraulic control unit	To set constant ploughing and sowing depth, compensating for load differences, bundling hay bales and even positioning the vehicle as it works its way up and down the field
Transmission control unit (power shuttle)	A power shuttle is an additional unit used in transmissions and is generally used in agricultural tractors. While the vehicle moves forward, the driver can pull a lever that makes it stop and go backwards at the same speed. Power shuttles are also known under trade name "Power Reverser".

Global Positioning System (GPS) (Fig 8.1.)	
Steering control system	To make perfectly straight parallel lines without the use of field markers or tracking devices.
Variable rate fertilizer applicator	To navigate the farm equipment for planting, harvesting, spraying of pesticides and herbicides, or applying fertilizers accurately.
	To run the tractor even in reduced visibility, dense fog, a moonless night, heavy rain or heavy dust.
Telematics	Tracking location, geo-fencing, health monitoring, usage, performance evaluation, theft, service request
Communication and Implements	
ISOBUS	The ISOBUS standard specifies a serial data network for control and communications on agricultural tractors and implements.
Sensor-based levelling (laser levelling)	To get uniform levelling in the entire area.

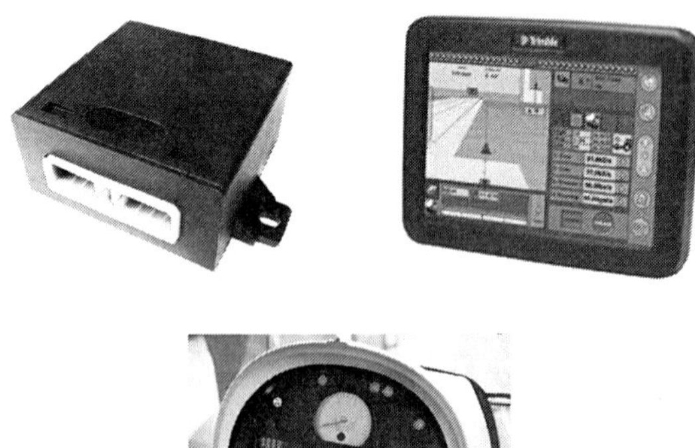

Fig 8.1. Features in new generation tractors
Source for fig: www. elensrl.it; www. avsagri.be; www. myfarmlife.com

The future tractors would have the following gear arrangement

Transmission	Tractor (hp)
12 x 12 transmission	60-90
24 x 24 transmission	90-130
Hydro static transmission	> 100
Power shift and Power shuttle	$\geq$75

SECTION 2
Agricultural Machinery

Permissions received from different institutions to reproduce photographs
Section 2: Agricultural Machinery

Figure No.	Title	Annexure (Permission letter and date)
Tractors and Farm Equipment Limited (TAFE)		
2.1	Mould board plough	TAFE products.
2.2	Disc plough	
2.3	Variants of rotary tiller	
2.4	Power harrow	
3.6	Potato planter	
6.2	Tractor mounted combine	
6.4	Potato digger	
Tamil Nadu Agricultũral University (TNAU)		
3.1	Direct rice seeder	Letter dated 20.04.2010
3.3	Air assisted seed drill	
6.3	Groundnut harvester	
7.1	Groundnut thresher	
Central Institute of Agricultural Engineering (CIAE)		
3.2	Animal drawn inclined plate planter	Email dated 22.02.2010
4.1	Wheel hoe	
5.1	Orchard sprayer	
8.1	Dal mill	
Indian Institute of Sugarcane Research (IISR)		
3.4	Ridger type sugarcane cutter planter	Email dated 29.03.2010
Indian Institute of Horticultural research (IIHR)		
3.5	Raised bed former cum seedling transplanter	Email dated 25.03.2010
Williames Tea Pty Ltd		
6.5	Selective tea harvester	Email dated 06.11.2014
Central Institute of Post Harvest Engineering and Technology (CIPHET)		
8.2	Mechanical pomegranate aril extractor	Email dated 17.03.2010
Others		
9.1	Different types of budding and grafting knives	Commercially available
12.1	Nomograph for seed drill	

1

Introduction

1. Introduction

Agricultural mechanization refers to usage of any improved tool, implement and machine by the farm workers to improve the efficiency of any operation to increase productivity. It also helps to reduce drudgery or stress which adversely affects human mental faculties leading to errors, imprecision and hazards and eventually loss of efficiency if done manually. In developed countries, however, mechanization is synonymous to automation. There is scope to mechanize every operation of crop production, post harvest and agro-processing.

1.1 Importance

Agriculture in India is unique in its characteristics, where over 250 different crops are cultivated in its varied agro-climatic regions, unlike 25 to 30 crops grown in many of the developed nations of the world.

A. Farm mechanization plays an important role in bringing about a significant improvement in agricultural productivity. The timeliness of operations has assumed greater significance in obtaining optimal yields from different crops. For instance, wheat is sown in Punjab up to the first fortnight of November. A delay beyond this period by every one week leads to reduction in the yield of about 150 kg/acre. This is also true for other crops. Timeliness is also important for farm operations such as weeding, irrigation, harvesting, threshing and marketing. These have to be done at an appropriate time or else the yield and farm income will be adversely affected. These jobs if manually dependent will involve lots of time and result in wastage of money.

B. The quality and precision of the operations are equally significant for realizing higher yields. The various operations such as land levelling, irrigation, sowing and planting, use of fertilizers, plant protection, harvesting and threshing need a high degree of precision to increase the efficiency of the inputs and reduce the losses. For example, sowing of the required quantity of seed at proper depth and uniform application of the given dose of fertilizer are only possible with the use of proper mechanical devices. However, when such operations are performed through indigenous methods, their efficiency is reduced.

C. The time taken to perform the sequence of operations is a factor that determines the cropping intensity. To ensure the timeliness of various operations, it is necessary to use mechanical equipment which has high output capacity. Mechanization has helped to increase the area under cultivation and the intensity of cropping.

D. Farm mechanization is gaining importance because the output of work done per hour is more than human labour. This also entails reduction in the total labour requirement. The ever-increasing wages for human labour, their non-availability during the time of need and rising cost of maintaining draught animals also speak in favour of mechanization.

1.2. Chronology of farm mechanization

Early agricultural mechanization in India was greatly influenced by the technological developments in England. Horse drawn and steam tractor operated implements have been imported during the later part of the nineteenth century. The horse drawn implements/ equipment imported from England were not suitable for bullocks used in India. Over a period of time these were modified to suit Indian draught animals.

Increase in the area under crops to meet the food requirement of the burgeoning population necessitated the usage of tractors for tillage and extensive usage of tractor powered or self propelled machineries.

Progress of agricultural mechanization in India

Pre-Green Revolution Era (1947-1965)	Green Revolution Era (1965-1975)	Post-Green Revolution Era (1975 onwards)
1. Farming by traditional methods 2. Low productivity (0.58 t/ha/year) 3. Increase in production was attributed to increase in the cultivated area	1. High yielding variety, fertilizer, chemicals inputs 2. Productivity increased (1.14 t/ha/year) 3. Need for timeliness in farm operations realized 4. Introduction of improved farm machinery	1. Use of scientific methods in the cultivation of crops 2. Raise in the productivity level 2.14 t/ha/year 3. Increase in the availability of farm

Pre-Green Revolution Era (1947-1965)	Green Revolution Era (1965-1975)	Post-Green Revolution Era (1975 onwards)
4. Farm power availability was about 0.27 kW/ha 5. Share of animal power sources was 98% to 88%	5. Farm power availability was about 0.47 kW/ha 6. Share of animal power sources decreased from 88% in 1965 to 62% in 1975	power to 1.2 kW/ha 4. Reduction in the share of animal power from 62% in 1975 to 24% in 1997. 5. More implements/ equipment and machinery put to use for all the farm operations

Farm mechanization is dependent mainly on the size of operational holding, land topography, availability of credit facilities and per hectare profitability which in turn is affected by per ha yield, cropping intensity, market prices, etc. The trend in growth of farm/agricultural machinery is presented (Table 1.1).

Table 1.1. Growth in farm/agricultural machinery (number in hundred)

Implement	1971-72	1976-77	1981-82	1986-87	1995-96	2000-01*	Growth rate (%)
Combine (tractor)	3.5	5.6	12	37	61.5	109	15.41
Harvester (self-propelled)	4.5	3.0	3.0	18.0	35.0	55.7	10.63
Mould board and disc plough	573	925	1429	2392	4004	5427	9.85
Disc harrow	556	1292	1892	3574	6751	9826	12.35
Cultivator	815	1766	3150	5956	11558	18444	13.57
Seed drill/ Seed ferti drill	246	640	1606	2777	7301	12609	17.38
Planter	85	244	305	443	643	798	9.06
Leveller	494	1201	4140	7008	11861	15912	13.72
Potato digger	-	-	569	878	1355	21813	22.00
Total threshers	2058	5041	10250	13638	19089	22104	9.08
Wheat thresher	1825	4278	8319	11599	16172	18763	8.90
Paddy thresher	136	575	1318	1148	1622	1767	9.86
Other threshers	97	188	613	891	1295	1574	10.71
Sugarcane crusher	872	1045	1208	1512	1892	2133	3.76
Electric pump	0.02	0.10	1.029	4.33	8.91	12.514	8.20
Diesel pump	0.083	0.230	1.546	3.101	4.659	5.940	4.89
Power sprayer/duster	-	-	0.045	0.124	0.200	0.311	7.10

*Estimated by Singh and Gajendra (1997)

1.3. Farm mechanization - Global scenario

Comparison of mechanization with other countries

Country	Farm power (kW/ha)	No. of tractors per 1000 ha	No. of combine harvester per 1000 ha
Japan	8.75	461.22	236.98
Italy	3.01	211.08	4.71
France	2.65	68.5	4.93
U.K.	2.5	88.34	8.3
Germany	2.35	79.817	11.41
India	1.36	15.75	0.026
Argentina	—	10.74	1.79
Brazil	—	13.66	0.915
China	—	6.98	2.53
Pakistan	—	16.47	0.08
Egypt	—	30.7	0.79

Source: FAO Yearbook 2003

Population engaged in agriculture and level of farm mechanization

Country	Population engaged (%)	Level of farm mechanization (%)
United States	2.4	95
Western Europe	3.9	95
Former Soviet Union	14.4	80
Argentina	9.4	75
Brazil	14.8	75
India	55.0	40
China	64.9	38
Africa	60	20

1.4. Mechanization of dryland agriculture: Status and constraints

Dryland agriculture constitutes about 60% of the total cultivated area in India and has a share of 42% of the total food production. Dryland area is spread over an area of 118 m ha of the gross cropped area. The average productivity is about 0.7 to 0.8 t/ha. One of the major causes of the poor productivity in drylands is the lack of mechanized operations and rainwater harvesting methods. Faster field operations assist directly or indirectly in conserving rain water and its effective use. Timeliness and precision in field operations are key factors in these areas. Small and marginal holdings are held by resource-poor farmers in the dryland regions. In recent times, human and animal resources have also been continuously dwindling in these regions. Conventional tools and equipment are used in these regions, but are largely inadequate to meet the requirement of precision farming. In view of these reasons, large-scale mechanization assumes greater significance.

Use of tractor power with suitable attachments will enhance timeliness in operation to utilize the available moisture for the establishment of crop and will further aid in its growth.

In drylands, most of the land holdings are small and scattered. Poor investment capacity of dryland farmers coupled with non-availability of quality implements and lack of adequate extension link are some of the constraints in the adoption of improved technology among farmers. The investment in farm machinery vis-à-vis their utilization on small farms works out to be high and therefore the majority of farmers give comparatively lower priority to purchase of agricultural machinery as compared to purchase of other inputs. Hence, both the government and NGO should get involved in farm mechanization work on a revenue-generating model so that it becomes sustainable. Important areas requiring mechanization include conservation of soil moisture by deep ploughing (chisel plough/sub-soiler) during summer in the farmers' field, construction of contour bunds with the consent of villagers to reduce runoff and soil loss and construction of common farm ponds and installation of water lifting devices.

1.5. Commercially exploited implements/equipment and machinery

Tractor mounted implements such as mould board ploughs, disc ploughs, cultivators and other crop-specific equipment are widely used for seed bed preparation. Seed drills and planters, both animal drawn and tractor mounted, have become popular. Mechanical transplanters for rice and vegetable crops are catching up with the farmers. Long handle tools and power weeders for weeding and interculture and manual and power-operated sprayers and dusters for application of chemicals have been commercialized. Cereal crop harvesters including various designs of vertical conveyor reaper windrowers and combine harvesters are being used on a large scale. Tractor mounted digger elevators are being used for groundnut and tuber crops. Spike-tooth, rasp-bar type and axial flow threshers are used in several major cereals. Besides, crop-specific threshers have been developed and commercialized for soybean, groundnut and sunflower.

1.6. Future requirement

The ensuing chapters on the availability of various implements/equipment/ machinery available for farm operations in many crops would reveal that many prototypes have been developed by several research institutes. But, only a few have been commercialized. The reasons include no refinement done to any of the prototypes after testing in the farmer's field for a sufficient length of time, reluctance among the small-scale industries to fabricate several prototypes because of low sales volume, etc. However, these can be overcome if there is a good linkage between research institutions and fabricators, public institutions take initiatives to popularize the usage of new farm equipment developed by them and custom hiring concept is encouraged amongst self help groups, NGO and village knowledge centres proposed to be established by the government.

The government's new initiative, Sub Mission on Agricultural Mechanization (SMAM), proposes to set up custom hiring centres and high-tech high-productive equipment centres in different states in the country through public–private partnership.

1.7. Government outlook

The government should ensure that

Agricultural mechanization leads to a sustainable increase in yields and cropping intensity with the objective of meeting the planned rate of growth in agricultural production.

Mechanization creates a worker-friendly environment. Workers do not need to fear that it would take away their livelihood.

Sufficient skill is imparted for operating many of the equipment at the farm level itself.

Banks extend loan to self help groups to purchase tractors. They in turn use the tractor for custom hiring.

1.8. Tools/implements/equipment/machinery covered

Tools, implements, equipment and machinery operated by hand, animals and tractors used for various field operations, *viz*., land preparation, sowing or planting, weeding, plant protection, harvest, threshing, post harvest and agro-processing, special horticultural tools, water lifting devices, calibration of seed drills, miscellaneous equipment and machinery are covered in the following chapters. An exhaustive list of products under each of the above-stated categories is provided. Besides, mention is also made of unique features of select products. This information can also be accessed from the web site *www.jfarmindia.com*

1.9. Disclaimer

Information provided in the following chapters does not represent or endorse the accuracy, completeness or reliability of any of the claims made by the respective institutes or commercial manufacturer for any of the prototype or commercial models developed by them. Similarly, information provided in this book disclaims warranties for all the products. Readers are advised to consult experts or evaluate individual products on their own before taking any decision to purchase any of the items based on the information provided in this book.

2

Land Preparation

2.1. What?

The soil is prepared to a fine tilth, which is referred to as land preparation or preparatory cultivation. This enables proper sowing of seeds, planting of seedlings and precise application of fertilizers to the crop. Various implements are used to prepare the land. The basic ones are ploughs, harrows, cultivators or tillers, and land levellers. These implements are manually operated or by animals or tractors. Various improvements have been made to increase the efficiency of each one of the implements by research workers. Many products are also available to suit the local regions and soil condition.

2.2. Plough

It has been the basic implement used for decades in agriculture to open up and invert the soil for initial cultivation. While inverting the soil, it also buries the weeds and the remains of the previous crops and allows them to decay. In the process of inverting, it also aerates the soil. Ploughs were initially used with animal power. Later on, many improved implements with heavy weights have been developed. These are drawn by tractors.

2.3. Harrow

The harrow is used to break up clods and lumps of soil left after the initial ploughing. This provides a fine finish to the soil structure and tilth. If the soil gets a good tilth, then, direct seeding or planting of crops can be taken up. During the process of harrowing, weeds are also removed.

There are different types of harrows, *viz*., disc and tine. Disc harrows are used for heavy work, like breaking of clods, while tined harrows are used to improve seed-bed condition before planting and to remove small weeds in the interspaces of newly emerging crop and to loosen up the inter-row soils to allow the water to soak and infiltrate into the subsoil.

2.4. Cultivator or tiller

Cultivator is a farm implement used for stirring and pulverizing the soil. In the process, it also aerates and loosens the soil. The cultivator is mainly used to prepare a proper seed bed for the crop to be planted. This implement can also be used in the inter-row spaces of crops up to one-month-old for removing weeds.

2.5. Puddler

The puddler is used in heavy clayey soil after impounding water. This is done to bring the soil to a muddy soft condition which helps water to stand on its surface. This type of soil preparation is required for planting rice. This is achieved by use of cage wheels fitted to a tractor.

2.6. Land levellers

These are used for levelling the field after ploughing and harrowing. It provides a proper surface gradient and helps to form irrigation channels and facilitate smooth flow of water.

2.7. Implements used in land preparation

2.7.1. Animal drawn

Plough, Bund former, Chisel plough, Dabra plough, Disc harrows, Harrow cum puddler, Helical blade puddler, Kapas ridger, Khargaon type hal, Levellers and land smootheners, Melur plough (TNAU), Mould board plough, Reversible mould board plough, MP iron wedge plough, Multipurpose tool frame (CIAE), Puddler, Ridger/ Lister plough, Cultivator, Bakhar (CIAE), Patela harrow (CIAE), Patela puddler (CIAE), Iron plough (TNAU), Improved plough, Multipurpose tool carrier (TNAU), Blade harrows (CIAE, PRKV), Lugged wheel puddler (CIAE), Harrow cum puddler (PAU)

2.7.2. Tractor drawn

Mould board plough (TAFE), Disc plough (TAFE), Subsoiler (TAFE), Duck foot cultivator, 9-tine / 11 - tine tiller (TAFE), Rigid tine cultivator, Heavy duty tiller, Rotary tiller (TAFE), Offset disc harrow (TAFE), Post-hole digger (TAFE), 3 Based ridger (TAFE), Terracer blade (TAFE), Peg type puddler, Spiked clod crusher (GBPUAT), Bed cum channel former (IGFRI), Cotton stalk uprooter (PAU), Laser leveler (PAU), Pulverizing roller attachment (PAU),

Stubble shaver (PAU), Broadbed former (TNAU), Channel former (TNAU), Combine tillage tool (TNAU), Power harrow (TAFE), Low draft chisel plough (TNAU), Para plough (TNAU), Trencher (TNAU)

2.7.3. Power operated

Terracer cum leveler (TNAU), Disc plough, Auger digger

2.8. Features of some commonly used land preparation implements

Implements	Features and efficiency
Animal drawn	
Country plough	Breaks the topsoil surface. It should be operated several times to obtain a satisfactory tilth. This cannot invert the soil. The depth of penetration is shallow
Patela harrow	Secondary tillage equipment used to crush clods, remove stubble and collect trash. Also used for levelling and smoothening of land before seeding. The overall dimension (l x w x h) of the implement is 1500 x 230 x 100 mm. The working width is 40 mm with a field capacity of 0.3 ha/hr.
Cultivator	Used for shallow ploughing, weeding and intercultural operations. Field capacity is 0.6-0.7 ha/day with a depth of operation of 10-15 cm.
Puddler	Provides well-homogenized, puddled tilth for transplanting paddy. Field capacity is 2-3 ha/day. The overall dimension (l x w x h) is 945 x 700 x 1060 mm. It has a working width of 700-800 mm and working depth of 100 mm. The weight of the implement is 40 kg
Bund former	Used to form ridges, bunds and mark boundaries of plots or fields. Helps to conserve rainwater and prevent runoff. Field capacity is 3-4 ha/day. The width of the implement can be adjusted between 750 mm and 1050 mm depending upon the requirement.
Levellers and land smootheners	Used to level the field after ploughing and harrowing. Certain models are provided with small tynes or steel hooks behind the levelling board to remove trash from fields. Wide levelling plank helps in compacting soil aggregates in the top layer which reduces evaporation of soil moisture. The width of the plank is 1.5-2.0 m. Field efficiency is 0.25-0.35 ha/hr.
Tractor operated	
Mould board plough (TAFE) (Fig 2.1)	Used for primary tillage. Share of the plough penetrates into the soil and makes a cut below the soil surface. Mould board lifts, pulverizes and inverts the soil. It has more under-beam and longitudinal clearances so that it can work in trashy conditions. It also has a special wear resistant bottom and bar point that can be extended / replaced after use. Two and three furrow variants are available. The maximum depth of operation is 300 mm. Operated using a 40 - 50 hp tractor.

Reversible mould board plough	Used for primary tillage in virgin land, fields left unploughed for long periods, fields with too hard/heavy clay soils and those heavily infested with deep-rooted unwanted plants, shrubs and weeds. The depth of cut is 200-250 mm and width is 150 to 225 mm. Tractor hydraulic operated and run with 55 hp tractor or more.
Disc plough (TAFE) (Fig 2.2.)	Basic tillage implement used in hard, dry, trashy, stony or stumpy land conditions and in soil where scouring is a major problem. The side draft is controlled by fully floating furrow wheel. Two and three bottom disc ploughs are available. The maximum depth of operation is 300 mm. Operated using a 35 - 50 hp tractor.
Offset disc harrow	Used in orchards, plantations and vineyards. Have two gangs of discs mounted one behind the other. The discs on the front gang throw the soil outward and the rear gang inward. Field efficiency is 2.5 ha/day. Operated using a 35 hp tractor or more.
Chisel plough/ Subsoiler	Used to break the hard pan below the normal ploughing depth to facilitate percolation of rainwater. Stimulates deep root growth and helps the crops to withstand drought. Aerates the soil with negligible damage to the surface. Also called sub-soiler. More useful in orchards. Breaks hard pan up to a depth of 535 mm. Operated using a 35 hp tractor or more.
9-tine/11-tine tiller	Multipurpose tillage implement used for preparing the land to sow or plant. It also helps to obtain good tilth that is a pre-requisite to form seed beds quickly and economically. Spring and rigid types are available. Field capacity is 0.35-0.5 ha/hr with a working width of 2100-3000 mm and working depth of 140-170 mm. Operated using a 35 hp tractor.
Rotary tiller (TAFE) (Fig 2.3.)	Used for final land preparation before forming a seed bed to take up sowing in a single pass. Stubbles and roots are completely cut and mixed with soil. It is also suitable for incorporating straw and green manure in the field. The implement has 6-8 flanges with 6 blades per flange. The working width is 1000-2000 mm. Different variants of rotary tiller with 20, 28, 32, 36, 42 & 48 blades are available. Operated using 35 - 50 hp tractor.
Power harrow (TAFE) (Fig 2.4.)	Suitable for garden land conditions. Used for secondary tillage and aerating the soil. Power is taken from tractor PTO. Operated by 40 hp tractor
Terracer blade	Ideal heavy duty implement with strength for scraping, grading, levelling and backfilling. Most suitable for irrigation, terrace work and general clean-up of fields. It has a cutting or scraping blade and thick curved sheet closed from the sides to form a bucket. Operated using a 35 hp tractor.
Post-hole digger	Used for making pits to plant saplings in orchards and erection of fencing/communication posts upto a depth of 90 cm. It can be fitted with augers of varying diameters, *viz.,* 9", 12", 18" and 24". Operation is faster, safer and most economical. Operated using a 35 hp tractor or more.

Power operated	
Terracer cum leveler	Suitable for land levelling, terracing and forming bunds. The overall dimensions (l x w x h) are 1000 x 780 x 580 mm and can be operated using a 8-10 hp power tiller. The field capacity is 1.2 to 1.5 ha/hr.

Two furrow MB plough   Three furrow MB plough

Fig 2.1: Mould board plough - TAFE

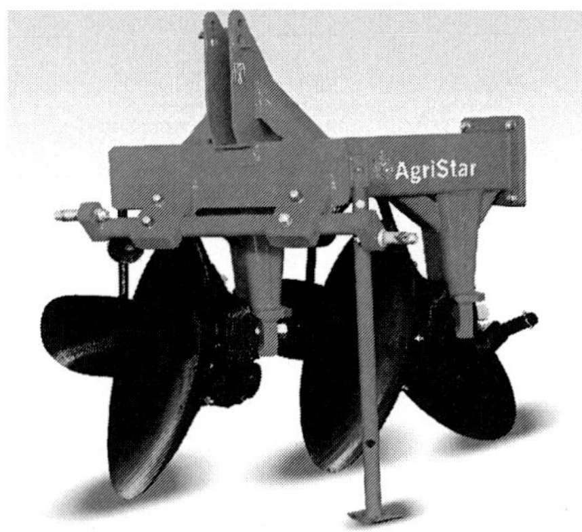

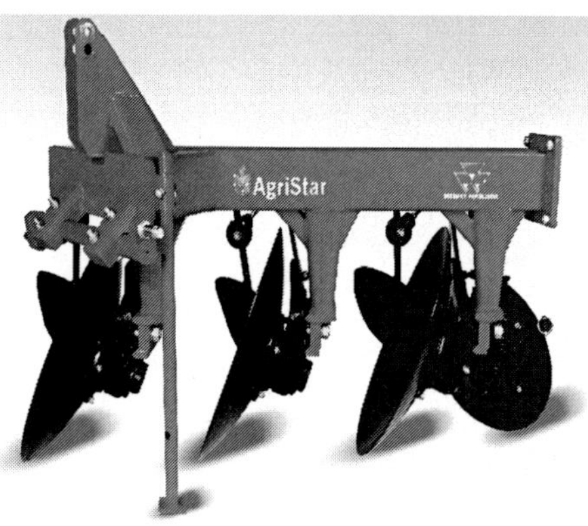

Two bottom disc plough   Three bottom disc plough

Fig. 2.2: Disc plough - TAFE

Rotary tiller - 30 blades

Rotary tiller - 48 blades

Rotary tiller - 54 blades

Fig 2.3: Variants of rotary tiller - TAFE

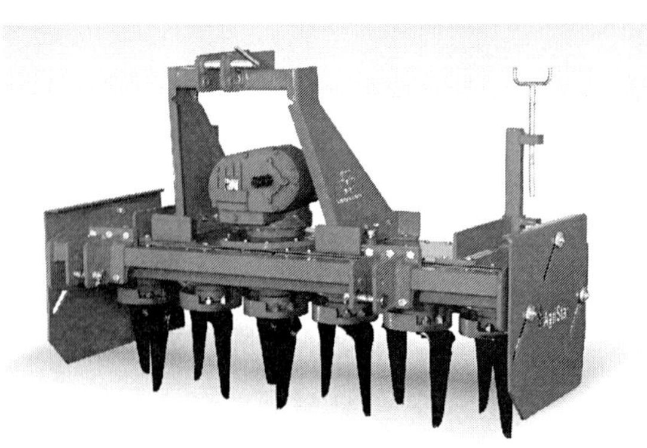

Fig 2.4: Power harrow - TAFE

3

Sowing and Planting Equipment

3.1. What?

The basic objective of sowing is to place the seed in rows at the desired depth, to maintain proper spacing between seed to seed, cover the seeds with soil and provide proper compaction over the seed if required. Precise sowing/planting helps in better crop stand. This helps to achieve optimum yields. Traditional sowing methods include broadcasting manually, opening furrows by a country plough and dropping seeds by hand (i.e.) dibbling. In the present day, improved seed-cum-fertilizer drills are used for sowing and fertilizer application. These are provided with seed and fertilizer boxes, metering mechanism, furrow openers, covering devices, frame, ground drive system and controls for variation of seed and fertilizer rates.

3.2. Types of sowing/planting equipment

Sowing equipment include seeders and transplanters. Various types of sowing equipment, *viz.*, manually operated, animal drawn, power operated and tractor drawn are used throughout India.

3.2.1. Seed drills

Seeders are available for sowing seeds of various crops. Some seeders have been designed with adjustable metering mechanism which makes them versatile for sowing two or three kinds of seeds. For example, a multicrop planter developed by CIAE is used for sowing garlic, maize, moong, peas and groundnut. Similarly, the institute has developed a small seed sowing drill which is used for sowing small seeds such as bajra, rapeseed and mustard.

3.2.2. Seed cum fertilizer drills

To save on time and labour, seed drills with inbuilt fertilizer drills are widely used. In certain places, along with fertilizer drill granular pesticide application is also made, e.g. sugarcane cutter planter with discs developed by IISR.

3.2.3. Zero till drills

These can directly drill seed and fertilizer without seed bed preparation and have five inverted 'T'-type furrow openers. A compaction roller is provided for closing the slit created by furrow openers after the placement of seed and fertilizers. These are suitable for drilling seeds of wheat, barley, lentil, chickpea and pea after rice harvest.

3.2.4. Transplanters

Transplanters are available for crops such as rice, brinjal, chilli, tomato, cauliflower and cabbage. These have a provision for one or two operators to sit and place the seedlings on the moving chain, which carries the seedlings to the planting mechanism. Two inclined wheels having a small flange at the outer periphery help in the placement of the seedlings and cover the soil.

All these seed drills should be calibrated to discharge the required seed rate following the specific instructions provided by the manufacturers. Calibration of seed drill is dealt broadly in Chapter 12 of this book.

3.3. Criteria for selecting sowing and planting equipment

Different designs of improved seed drills/planters have been developed for sowing of different crops. Seed drills vary mainly in the type of seed metering mechanism and furrow openers. Hence, seed drills need to be selected to suit the crop and soil conditions.

3.3.1. Based on seed metering mechanism

- For small seeds like rapeseed–mustard seed, drills with vertical roller with cells, inclined seed plate with cells or small grooved fluted roller metering system is suited.
- For medium seeds such as wheat, soybean, safflower and linseed, seed drills with standard fluted rollers are recommended.
- For bold seeds like groundnut and castor, planters with inclined cell plate or cup feed type metering system are recommended.

3.3.2. Based on the type of furrow openers

Furrow openers should be selected according to the type of soil and depth of seed placement.

- For trashy, stony and light to medium soils, shovel type openers which are capable of placing seeds between 50 and 100 mm depth are used.
- Small shoe type openers are also used for shallow placement of seeds in dry farming areas.
- Shoe type openers with single or twin boots are used for sowing in heavy and medium soils for seed placement at 20 to 70 mm depth.

3.4. Equipment developed by various ICAR institutes and state agricultural universities

3.4.1. Manual

Direct rice seeder (TNAU), Mat type nursery sowing seeder (PAU), 6-row rice transplanter (PAU), Mustard drill (PAU), Rice transplanter (CIAE), Rice cum green manure seeder (TNAU), Seed treating drum (APAU), Seed and fertilizer drill (CIAE), Multicrop planter (CIAE), Oilseed drill (PAU), Rotary dibbler (CIAE), Gorru (TNAU), Broadcaster (TNAU), Single row garlic planter (PAU)

3.4.2. Animal drawn

Seed cum fertilizer drill (ANGRAU), Birsa seed cum fertilizer drill (BAU), Seed cum fertilizer drill (CCSHAU), 3 -row seed cum fertilizer drill (CIAE), 3-row low cost seed-cum-fertilizer drill (CIAE), Low cost seed drill (CIAE), Inclined plate planter (CIAE), Small seed sowing drill (CIAE), Tool frame (CIAE), Planter (CIAE), Jyoti planter (MPKV), Seed cum fertilizer drill/ planter (PAU), Seed drill (Boram dev seed drill), Seed cum fertilizer drill (Hoshangabad duphan), Seed cum fertilizer drill (Khurai type duphan), Seed cum fertilizer drill (Nimad-type tiphan), Single tine seed cum ferti drill (Raipur type), Planter (TNAU), 2-row rape/ mustard seed cum fertilizer drill (CIAE), Mustard seed drill (CCSHAU), Rapeseed-mustard seed drill (PAU), 3-row groundnut planter (CIAE), 3-row groundnut planter (TNAU), Sugarcane planter (IISR), Potato planter

3.4.3. Tractor operated

Multipurpose ferti-seed-planter (APAU), Pneumatic planter (for bold seeds) (CIAE), Modular planter (CIAE), Inclined plate planter (CIAE), Seed cum fertilizer cum planter (PAU), Vertical disc planter (PAU), Bed planter (PAU),

Seed-cum-fertilizer drill (PAU), Seed drill, Air-assisted high-speed drilling attachment for small seeds (TNAU), Ridger seeder (TNAU), Improved broad bed cum seeder (TNAU), Basin lister cum seeder attachment to cultivator (TNAU), Mustard seed drill (GB Pant Univ), Groundnut planter (TNAU), Ridger planter for winter maize (PAU), Direct rice seeder (TNAU), Inclined plate type planter (CIAE), Sugarcane cutter planter (PAU), Semi-automatic sugarcane planter (PAU), Semi automatic sugarcane planter (IISR), Mustard seed drill attachment to semi automatic sugarcane planter (IISR), Sugarcane cutter planter with discs (IISR), Ridger type sugarcane cutter planter (IISR), Automatic sugarcane cutter planter (IISR), Cotton ridger seeder (TNAU), Inclined cell type planter (CCSHAU), Till planter (PAU), Strip till drill, Zero till drill, Potato planter (TAFE), Automatic planter cum intercultivator (CIAE), Semi automatic potato planter (PAU), Pneumatic planter (for light vegetable seeds) (CIAE), Vegetable transplanter (CIAE), Raised bed former cum vegetable transplanter (IIHR), Countersinking attachment (KAU)

3.4.4. Power operated

Seed cum fertilizer drill (CIAE), Zero till drill machine (CIAE), Till plant machine (CIAE), Seeder (TNAU), Rice transplanter (single wheel driven and four wheel, riding type), Paddy transplanter (PAU), Cotton seed delinting machine (TNAU)

3.4.5. Self-propelled

Self propelled drum seeder (CRRI), Rice transplanter (Daedong DP-480)

3.5. Features of some commonly used sowing equipment

Equipment	Features
Manual	
Direct rice seeder (TNAU) (Fig 3.1)	Ensures uniform seed distribution and helps in maintaining uniform plant population per unit area. Power is transmitted from the ground wheel to the drum shaft which aids in seed distribution. The overall dimension (l x w x h) is 2 x 1.5 x 0.65 m with a weight of 10 kg. Field capacity is 1.1 ha/day with an efficiency of 65%.
Mat type nursery sowing seeder (PAU)	Uniformly spreads pre-germinated paddy seeds over the soil-filled frames. The dimension of the seeder (l x w x h) is 1380 x 1200 x 280 mm. It has a cylindrical roller of 190 mm diameter with 10 mm diameter holes on it. The roller is with 4 agitators which help in delivering the seeds uniformly.

6-row rice transplanter (PAU)	Transplants mat-type rice seedlings in puddled and levelled soil. Its overall dimension (l x w x h) is 146 x 133 x 53 cm and it can plant 6 rows simultaneously. Field capacity is 0.25 ha/man day.
Seed and fertilizer drill (CIAE)	Suited for sowing soybean, maize, pigeonpea, sorghum, green gram, bengal gram and wheat along with the fertilizer. The field capacity is 0.05 ha/hr with an efficiency of 85-90%. One operator pulls the implement and the other guides the implement.
Animal drawn	
3 -row seed cum fertilizer drill (CIAE)	Suitable for sowing wheat, gram, sorghum, soybean, lentil, pea, sunflower and safflower along with the fertilizer. The overall dimension is 1000 x 1000 x 780 mm with 50 kg weight. It has a field capacity of 0.1-0.28 ha/hr with an efficiency of 65-70%
Small seed sowing drill (CIAE)	Used for drilling small seeds such as bajra, rapeseed and mustard. The overall dimension (l x w x h) is 1250 x 1110 x 950 mm with 50 kg weight. Field capacity is 0.12-0.16 ha/hr with 65-70% efficiency
Inclined plate planter (CIAE) (Fig 3.2)	Suitable for planting groundnut, maize, pigeonpea, sorghum and other oilseeds. It completes sowing of three rows in a single pass. The overall dimension (l x w x h) is 4200 x 1300 x 900 mm. It covers 0.12-0.15 ha/hr with a field efficiency of 60-65%.
3-row groundnut planter (CIAE)	Individual seed metering units with separate boxes provided in the planter can be used for sowing intercrops. Suited for sowing groundnut, soybean, safflower, mustard and sesame. It has a shoe type furrow opener for placing the seeds at the desired depth. Field capacity is 0.16-0.24 ha/hr.
Sugarcane planter (IISR)	Used for planting straight cane setts and application of granular fertilizer. Furrow opening, placement of setts, covering of the setts and application of granular fertilizer are completed in a single pass. Field capacity is 0.1 ha/hr with 30% efficiency. The metering mechanism gets its drive from the ground wheel and a set of chain and sprockets. Saves 65% on time, 55% on cost and 60% on labour compared to conventional manual cutting and closing the setts with a country plough.
Potato planter	Makes ridges and places the tubers on the ridges and covers the soil in a single pass. The planter has seed box to place the tubers.The overall dimension (l x w x h) is 1150 x 1000 x 950 mm with 900 mm coverage width. Field capacity is 0.1 ha/hr with 70% efficiency. 10% of missing is reported. Saves 85% on time, 60% on cost and 50% on labour compared to manual planting.

Tractor operated

Air-assisted high-speed drilling attachment for small seeds (TNAU) (Fig 3.3)	Suitable for sowing small seeds such as ragi, maize, sesame and mustard. It consists of a seed hopper, feed regulator, air blower, seed feeding mechanism and seed delivery system. Blower blows air through a vertical distributor tube to the distributor head for carrying the seeds to the delivery system. The unit is mounted on a 9 tined cultivator and operated by 35-45 hp tractor. It has a field capacity of 5 ha/day. It saves 55% on cost and 92% on time compared to conventional sowing.
Groundnut planter (TNAU)	Suitable for sowing groundnut and chickpea. It has a cup-type seed feed mechanism and a shovel type furrow opener for sowing the seeds. A square bar is provided at the back of the unit for covering the seeds. Field capacity is 0.4 ha/hr. Operated by 35 hp tractor.
Ridger type sugarcane cutter planter (IISR) (Fig 3.4)	Cuts the seed setts automatically to the desired size, opens the furrow, places the setts in the furrow, applies fertilizers, treats the sett and covers it. Row to row spacing can be adjusted from 75 to 90 cm with a furrow width of 40-45 cm. The overall dimension (l x w x h) of the unit is 204 x 162 x 211 cm. It has a field capacity of 0.25 ha/hr and an efficiency of 65%. Operated by 35 hp tractor or more. This provides 40% saving on cost compared to the conventional method of planting.
Inclined cell type planter (CCSHAU)	Suitable for sowing Bt cotton. Maintains 1.0 m spacing between plants and rows. Field capacity is 0.8 ha/hr.
Strip till drill	Tilling and sowing are done simultaneously. The seed drill has a rotary system with C-type blades. These make a 75 mm-wide strip in the front of every furrow opener. Thus with every row, 125 mm of the strip is left untilled. It has 9-11 furrow openers. Tills only 40% of the area and useful for sowing wheat after rice without prior seed bed preparation. Operated by 30-35 hp tractor.
Zero till drill	Used for sowing wheat in unprepared field after harvesting rice. The seed drill has an inverted T-type furrow openers (narrow shovels), which provide lower draft requirement and helps in easier penetration of soil. It has 9-13 furrow openers.
Automatic planter cum intercultivator (CIAE)	Suitable for planting potato tubers of 25-40 mm size at 25-40 cm spacing. It uses a round plate of 600 mm diameter with 12 fingers to pick the potato tubers. The overall dimension (l x w x h) is 900 x 1500 x 1350 mm with a working width of 600 mm and

	working depth of 100-200 mm. It is also used for intercultivation in row crops and earthing up operation. Field capacity is 0.4 ha/hr and efficiency is 80%. Operated by 30-35 hp tractor.
Raised bed former cum vegetable transplanter (IIHR) (Fig 3.5)	Has an inbuilt bed former and is used for planting vegetable seedlings and tubers. It has two ridgers for forming the beds and punch wheels for making pits on the bed. The operator sits on the frame and drops the seedling manually in each pit. The seedlings are kept in protray over the stand. The pit is closed by the shovels. The number of punches on the wheel varies from 4 to 8 depending upon the plant spacing. Requires 30-day-old seedlings of 10 cm height for transplanting. Provision for fitting a water tanker and the necessary piping for watering the transplanted seedlings are also provided as per customer requirement. Field capacity varies from 0.11 to 0.17 ha/hr depending upon the spacing and type of seedling planted.
Potato planter (TAFE) (Fig 3.6.)	Two row planter with a provision to adjust the spacing between plants, rows and depth of planting. Rigid and concave mould boards are used for forming the ridges. Completes planting and fertilizer application in a single pass. Operated by 40-50 hp tractor.
Power operated	
Seed cum fertilizer drill (CIAE)	Suitable for sowing wheat, sorghum, soybean and chickpea in medium and heavy soils. It has a depth control gauge wheel which helps in placing the seeds at desired depths. Overall dimension (l x w x h) is 1650 x 1510 x 1130 mm. Field capacity is 0.2-0.25 ha/hr with 50-60% efficiency. It can be operated using 8-10 hp power tiller.
Zero till drill machine (CIAE)	Directly drills seed and fertilizer without seed bed preparation. It has five inverted 'T'-type furrow openers and a compaction roller which close the seed after sowing. Suitable for drilling seeds of wheat, barley, lentil, chickpea and pea after rice harvest. Field capacity is 0.15 ha/hr with an efficiency of 65%. It can be operated using 8-10 hp power tiller.
Self-propelled	
Self propelled drum seeder (CRRI)	Suitable for sowing germinated paddy seeds in low land and hilly terrain. Field capacity is 0.235 ha/hr
Rice transplanter (Daedong DP-480)	Plants six rows in a single pass. The riding type transplanter has a field efficiency of 0.25 ha/hr. The row to row spacing is 30 cm, while plant to plant spacing can be adjusted between 14 and 16 cm. Operated by petrol and the weight of the transplanter is 650 kg.

Fig 3.1. Direct rice seeder – TNAU

Fig 3.2. Animal drawn inclined plate planter – CIAE

Fig 3.3. Air assisted seed drill – TNAU

Fig 3.4. Ridger type sugarcane cutter planter – IISR

Fig 3.5. Raised bed former cum seedling transplanter – IIHR

Fig 3.6. Potato planter - TAFE

4

Weeding and Intercultivation

4.1. Need for weeding

Weeding is the critical operation to be carried out to obtain a good crop. Weeds are plants which grow in places where they are not wanted. They grow in the fields competing with crops for water, soil nutrients, light and space and thus reduce crop yields. They also harbor insect pests and micro-organisms. Certain weeds release the inhibitors or poisonous substances into the soil which are harmful to plants, human beings and livestocks. They increase the expenditure on labour and equipment, render harvesting difficult and reduce the quality and marketability of agricultural produce. When grown in water channels, they block and or reduce the flow of water.

The weed seeds germinate earlier and grow faster than the crop plant. This phenomenon is more pronounced when coupled with favourable environment like fine soil tilth and the soil receiving adequate water and fertilizers. Weeds flower earlier and produce seeds profusely ahead of the crop. So, these seeds get mixed up with the harvested crop and are distributed to the next season or transported to other places along with the harvested produce.

4.2. Earthing up

Earthing up refers to piling up or heaping soil around the base of the plant. This protects the plant from lodging and exposure of roots on the surface of soil. In certain instances, earthing up becomes more crucial as in groundnut. Earthing up in this crop ensures proper peg formation and thereby increases the number of pods per plant. In cotton, earthing up prevents the entry and damage caused by stem weevil. In potato, it helps in the production of increased number and size of tubers. Earthing up can be done manually using a spade or with implements drawn by tractor.

4.3. Types of weeders

To remove weeds and at the same time use them as soil mulch, several weeders have been developed by different agricultural institutions. Use of these weeders reduces human drudgery, labour and cost as compared to hand weeding. Weeders are broadly classified as manually operated, animal drawn, tractor operated and power operated.

All the manually operated hoe type weeders work on the basis of push and pull type. These are fitted with various types of small pegs. By forward and reverse movement, the pegs will uproot and cut the weeds.

Animal drawn, tractor and power-operated weeders have tynes with shovel blades which penetrate into the soil and uproot weeds.

4.4. Implements used for weeding and intercultivation

4.4.1. Manual

Sugarcane earthing hoe, Hand wheel hoe (Naini type) (CIAE), Peg type dryland weeder (CIAE), 3 tined hand hoe/ grubber (CIAE), Wheel hoe (CIAE, PAU), Cono Weeder (TNAU), Improved dryland weeder (TNAU), Star wheel dryland weeder (TNAU), V blade hand hoe (MP State Agro), Three tined hand cultivator (CIAE)

4.4.2. Animal drawn

Blade harrow (MP single dora), Blade harrow (MP double dora), Peg weeder, Peg tooth lever harrow, Triangular harrow, Three tined Bardoli hoe (ATRC), Multipurpose hoe (MPKV), Star weeder (APAU)

4.4.3. Tractor drawn

Stubble shaver (IISR), Weeding cum earthing up equipment (TNAU), Earthing up cum interculture equipment (GBPUAT), Multirow rotary weeder (TNAU)

4.4.4. Power operated

Portable powered cultivator, Self propelled power weeder (CIAE), Sweep cultivator (CIAE), Power rotary weeder (TNAU), Cotton stalk puller, Basin opener (CPCRI)

4.5. Features of some commonly used weeders and intercultivators

Implement	Features
Manual	
Sugarcane earthing hoe	Used for earthing up in sugarcane planted with row spacing of 90-120 cm. Suitable for loosening the soil between the rows and moving it towards the standing crop for earthing. It has a field capacity of 0.14 ha/hr with an efficiency of 77%.
Hand wheel hoe (Naini type) (CIAE)	Suitable for weeding in light soils for crops sown in rows. The wheel hoe is pushed front and back to uproot the weeds. Field capacity is 0.3-0.4 ha/day.
Peg type dryland weeder	Suitable for weeding in upland row crops. The diamond-shaped pegs with convex or 'V'-shaped blade penetrate into the soil and the rolling action pulverizes the soil. The weeder is repeatedly pushed and pulled in between the crop rows in standing position. It has an output capacity of 0.12 ha/man day with a working width of 20 cm.
Wheel hoe (CIAE) (Fig 4.1)	Has one or two wheels with 200-600 mm diameter fitted to a long handle. Different types of soil working tools such as straight blade, reversible blades, sweeps, V-blade, tine cultivator, pronged hoe, miniature furrower and spike harrow can be fitted to this depending upon the soil condition. Operated by repeated push–pull action which cuts/uproots the weeds in between the crop rows. It can penetrate up to 6 cm depth.
Cono weeder (TNAU)	Useful for weeding between two rows of rice crop, as it does not sink in the puddle. The weeder consists of two rotors, float, frame and handle. The orientation of rotors creates a back and forth movement in the top 3 cm of soil. The dimension (w x h) of the weeder is 370 x 1400 mm. It has a field capacity of 0.18 ha/day.
Star wheel dryland weeder (TNAU)	Manual push type weeder suitable for weeding upland row crops. The star wheels break the soil crust. The blade cuts and uproots the weeds. The blade can be adjusted to vary the depth of working. The overall dimension is 174 x 23 x162.5 cm. It has a working width of 15 cm and covers 0.1-0.2 ha/man day.
Animal drawn	
Blade harrow (MP double dora)	Used for weeding in cotton, maize and sorghum. The cutting width of the blade is 350 mm. It has a field capacity of 0.8-1.0 ha/day.

Peg weeder	Used for weeding in crops sown by broadcasting and breaking of soil crust. The pegs are bent at an appropriate angle for penetration in the soil. Overall dimension (l x w x h) is 1300 x 480 x 870 mm with a working width of 1350 mm. It has a field capacity of 2-2.5 ha/day.
Triangular harrow	Used for secondary tillage, intercultivation, breaking of soil crust and pulverization of soil. It consists of a triangular frame with iron sections and square pegs. Due to the triangular construction, full coverage is achieved and a fine tilth is produced in a single pass. The overall dimension (l x w x h) of the implement is 1850 x 500 x 200 mm with 35 pegs.
Multipurpose hoe (MPKV)	Suitable for weeding and earthing up in upland row crops. It has three sweep type hoes. On its rear, a V-shaped attachment is provided to level the soil and move the soil sideways. The overall dimension is 280 x 105 x 100 cm. It covers 0.15-0.2 ha/hr with a weeding efficiency of 75-85%.
Tractor drawn	
Stubble shaver	Used for removal of rice and wheat stubbles after harvest of the crop by the combine harvester. Also used in shaving stubbles in the harvested sugarcane field to raise a ratoon crop. It is tractor PTO-operated and consists of horizontally rotating blades, that chop residues or stubbles into fine pieces by impact. The power source is 30 hp tractor. It has a cutting width of 140mm.
Weeding cum earthing up equipment (TNAU)	Suitable for weeding and intercultural operations in between row crops. It has three sweep type blades fixed to the ridger frame for performing the weeding operation and three ridger bottom fitted behind the sweep blade for earthing up. Weeding and earthing up operations are performed in a single pass. The sweep blades and ridger bottoms are adjustable to operate at 45, 60, 75 and 90 cm. It has an efficiency of 61% and covers 1.6 ha/day. The power source is 35 hp tractor.
Multirow rotary weeder (TNAU)	Suitable for weeding and intercultural operations in between rows of crops such as sugarcane, cotton and maize. The overall dimension (l x w x h) is 400 x 636 x 1665 mm. It has a field capacity of 2 ha/day with a weeding efficiency of 70%. It helps in saving 73% on the cost and 82% on time. Operated by a 35 hp tractor.
Power operated	
Self propelled power weeder (CIAE)	Suitable for weeding and intercultural operations in upland row crops like groundnut, maize, soybean and pigeon pea sown in 30 cm row spacing. The tynes can

be raised or lowered depending upon the depth of operation. The spacing between sweeps is easily adjusted by sliding the tynes to suit the crop row spacing. The overall dimensions (l x w x h) are 2300 x 1000 x 850 mm. It is powered with 3 hp petrol engine for start and further run is by kerosene engine. It has a field capacity of 0.15 ha/hr with a weeding efficiency of 65-70%.

Power rotary weeder (TNAU)	Suitable for mechanical removal of weeds in cotton, tapioca, sugarcane, maize, pulses, tomato and in orchards where the row spacing is more than 45 cm. The overall dimension (l x w x h) is 2400 x 1750 x 1100 mm. Operated by 8.28 hp diesel engine. Field capacity is 1-1.2 ha/day.
Others	
Plastic mulch laying machine	Used for laying of plastic films of width 500-1500 mm as mulch. It consists of two concave discs, pneumatic press wheels, pneumatic transport wheels and scrapers for covering plastic film roll edges with soil. It has a field capacity of 0.12-0.18 ha/hr with an efficiency of 55.4-62.5%. Used for efficient moisture conservation, weed control, maintenance of soil structure and microenvironment.

Fig 4.1. Wheel hoe – CIAE

5

Spraying Equipment

5.1. What?

Sprayer is equipment used for applying insecticides, fungicides, herbicides and in some cases fertilizers as foliar sprays to agricultural crops. Sprayers range in size from man-portable units (backpacks) to Self propelled units with boom mounts covering 60-150 feet length. Sprayers include the ones used for spraying liquid formulations, dusters for dusting dust formulations and granular applicators for delivering granules.

5.2. Principle of working

Sprayers work based on Bernoulli's principle. Basically, the sprayers have a storage tank, a plunger, lance and a nozzle. When the plunger is pushed in, the air flows at a high velocity through the nozzle. The flow of air at high velocity creates a region of low pressure just above the lance. The higher atmospheric pressure acts on the surface of the liquid insecticide causing it to rise up the lance and the insecticide leaves out through the nozzle as a fine spray.

Sprayers are fitted with hollow cone nozzles for spraying pesticides and with flat fan nozzles for herbicides.

5.3. Classification of sprayers

Sprayers are classified as high volume (HV), low volume (LV) and ultralow volume (ULV), according to the total volume of liquid applied per unit of ground area. Initially, high-volume spraying technique has been used for pesticide application, but with the advent of new pesticides the trend is to use the least amount of carrier or diluents' liquid. In spraying, the optimum droplet size differs for different types of application. Fine droplets are required for pesticides and bigger size droplets for application of herbicides. The greater the number of fine droplets produced by the sprayer, the better will be the deposition on the target area. The size of the droplet is important as it affects the drift and penetration distance of droplets towards the target surface.

Quantity of water used in sprayers

Type of sprayer	Volume of water (l/ha)
High volume	500 - 1000 depending on the crop and spread of foliage
Low volume	200 - 500 depending on the crop and spread of foliage
Ultralow volume	20 - 80 depending on the crop and spread of foliage

5.4. Types of sprayers

5.4.1. High-volume sprayers

Foot sprayer, Knapsack sprayer, Rocker sprayer, Tractor mounted high-volume sprayer, Twin Knapsack with boom

5.4.2. Low-volume and Ultralow-volume sprayers

Mist blower cum duster, Low volume knapsack spinning disc sprayer, Motorized knapsack sprayer cum duster (TAFE), ULV

5.4.3. Others

Aeroblast sprayer, Blower sprayer, Front mounted self propelled boom sprayer, Hand compression sprayer, Hand rotary duster, Stirrup pump sprayer, Tractor mounted sprayer for cotton (PAU), Orchard sprayer (CIAE), Power sprayer, Self propelled high clearance sprayer, Self propelled light weight boom sprayer, Fogging machine, Tree duster, Tree sprayer, Telescopic pipe assembly (CPCRI), Areca sprayer (TNAU), Coconut tree sprayer (TNAU)

5.5. Calibration of sprayers

A sprayer needs to be calibrated to have the desired output of spray fluid per unit time. The three most important points to be considered for calibrating a sprayer are

1. The walking speed of the operator expressed in kilometre per hour (kph)
2. The output per minute of the sprayer expressed in litre or a part thereof
3. The width of each pass of the sprayer, commonly known as the swath width or spray width expressed in metre or part thereof.

Procedure of calibrating the sprayer

1. Assessing the walking speed

- Measure a distance of 100 metres.
- Build up the pressure, start spraying by holding the lance at knee height and walk at a steady walking pace.
- Make 5 passes each of 100 m distance and note down the time taken to cover each of the 100 m distance. Calculate the average and work out the speed in mps. This can be converted to kph using the formula 1 mps = 3.6 kph.

For example, if 75 seconds is taken to cover 100 m distance, then in one second 1.33 m is covered. This is equivalent to 4.8 kph (1.33 x 3.6 = 4.788).

2. Measuring the sprayer output

Pressurize the sprayer fully, release the cutoff trigger and hold the nozzle into a metric calibrated measuring jar for one minute. Keep pressurizing the sprayer evenly.

Note the amount of water collected in the measuring jar; this will give you the output of the sprayer.

3. The swath width or spray width

Pressurize the sprayer at constant speed. Hold the nozzle at knee height and start spraying on a clean dry track, so that the wet patch of spray width is visible.

Spray for one minute and measure the width of the wet patch at three different points within the one metre distance. Ensure that the wet patch is marked and measured to have an accurate value.

A calibrated sprayer will discharge the same amount of spray volume each time.

5.6. Features of some commonly used sprayers

Sprayer	Features
High volume	
Knapsack sprayer	Provided with an inbuilt storage tank. Uniform pressure can be maintained by continuously operating the pump. Recommended for spraying in row crops, small trees and

shrubs. Designs with varying tank capacity of 14, 16 and 18 litres are available. It has a swath width of 0.75-1.2 m. The field capacity is 0.05-0.1 ha/hr with an application rate of 250-400 l/ha. Operating pressure of the sprayer is 2-3 kg/cm^2.

Rocker sprayer

The sprayer consists of a single- or double acting piston pump for developing high pressure. It does not have an inbuilt tank. The pump is operated with a long lever to and fro in a rocking motion to suck the spray fluid. The other person holds the lance and directs the spray fluid to the target surface. The complete assembly is mounted on a wooden board, which is held to the ground by the foot of the operator. Suitable for crops with wide row spacing and trees. It has a swath width of 1.5-2 m. The field capacity is 0.12-0.15 ha/hr with an application rate of 250-400 l/ha. The operating pressure of the sprayer is 3-4 kg/cm^2.

Foot sprayer

The pump barrel is mounted on a steel frame, which provides stability when placed on the ground. The sprayer does not have an inbuilt tank, so an additional container is required to store the spray fluid. The inlet pipe is placed in the storage container and one person continuously operates the pump by a foot lever. The other person directs the lance to the target. Suited for wide-spaced row crops. It has a swath width of 1.5-2m. The field capacity is 0.12-0.15 ha/hr with an application rate of 250-400 l/ha. The operating pressure of the sprayer is 3-4 kg/cm^2

Tractor mounted high-volume sprayer

Suitable for use when the crop is small and planted at a convenient spacing for the movement of the tractor. The complete sprayer is mounted on 3-point linkage of the tractor and the pump is operated by PTO power. Gives wide field coverage because up to 20 nozzles can be fitted. It has a tank capacity of 200-400 l and a swath width of 7-8 m. The field capacity is 0.75-1 ha/hr with an application rate of 250-400 l/ha. The operating pressure of the sprayer is 3-3.5 kg/cm^2.

Low volume and ULV

ULV

Also known as controlled droplet application (CDA) sprayers. ULV formulated chemicals can be sprayed with very low dilution or no dilution using these sprayers. Suitable for spraying in nurseries, vegetable crops, vineyards and other field crops. Operated using 12V dry or lead acid battery. One charge of battery lasts up to 15 hours of spraying. The sprayer has a high-speed spinning disc. When the spray droplet travels in the grooves of the disc, fitted to the motor, it gets fragmented into very fine droplets. These droplets are thrown out due to the centrifugal force. It has a tank capacity of 1-2.5 l and a swath width of 1.2-1.5 m. The field capacity is 0.15-0.2 ha/hr. Operated at atmospheric pressure.

Others

Front mounted self-propelled boom sprayer	The pump and boom with 14 nozzles are mounted on a stand, which is attached in the front. The power for operation is derived from a 5.5hp diesel engine. Used for uniform spraying in horticultural crops and other crops such as cotton, maize and groundnut. It has a tank capacity of 50 l. The field capacity is 0.6 ha/hr with an efficiency of 65%. Covers a width of 6500 mm.
Hand compression sprayer	The inbuilt tank is filled to three-fourths of its capacity and pressurized by hand plunger pump to get the discharge. For maintaining proper atomization of the spray liquid, the tank requires frequent pressurization. Used in nurseries, vegetable gardens and flower crops. The field capacity is 0.4 l/ha. It works at a pressure of 3-4 kg/cm^2.
Hand rotary duster	The duster has a mechanical agitator to agitate the chemical in the hopper and prevent clogging. The fan/blower is enclosed in the casing and is rotated with the handle through the gearbox. The discharge pipe is directed towards the target to deliver the dust formulation. Available in shoulder-mounted and belly-mounted models. The hopper holds 5 kg of chemical. It has a field capacity of 0.6 ha/day.
Orchard sprayer (Fig 5.1)	Has a horizontal triplex piston pump, hydraulically agitated chemical tank and boom with turbo nozzles that are operated with a pressure of 9-18 kg/cm^2. It generates droplets of 100-150 micron size. Recommended for spraying for fruit trees such as pomegranate, orange and sweet lime and in vineyards. The power for the sprayer is tapped from a 12-14 hp power tiller with a fuel consumption of 1.15 l/hr. The sprayer has a tank capacity of 200 l and fitted with 6 nozzles. It has a field capacity of 0.4-0.7 ha/hr.
Power sprayer	Uses high pressure and high discharge for covering a large area. Operated by auxiliary engines or electric motors. Suitable for spraying in orchards, tea, coffee plantations, rubber plantations, vineyards and tall trees (up to15 m). It utilizes 1-3 hp petrol engine as a power source and can discharge up to 25 l/min. The field capacity is 0.2-0.3 ha/hr.
Tree sprayer	Consists of a 3 hp, 4- stroke petrol/ kerosene engine. The tank is filled with spray liquid and the engine is started by cranking with the rope to drive the fan and the rotary pump. Flow rate is adjusted by a control valve. The sprayer is placed under the tree and manually moved around it to complete the spraying operation. Suitable for spraying tall trees up to 25 m height. Field capacity is 3-4 ha/day.
Telescopic pipe assembly (CPCRI)	Used for spraying coconut and arecanut trees up to a height of 42 feet. This can be operated either by power or hand.

Fogging machine	Used to vaporize pesticides in the form of fog for killing flying insects. The spray liquid is vaporized, which condenses on contact with air. This is more commonly used in mosquito control programs. It utilizes LPG or petroleum-based products as a fuel source.

Fig 5.1. Orchard sprayer – CIAE

6

Harvest

6.1. What?

Harvest is the process of gathering the end produce from mature crops in the fields. Harvest marks the end of the growing season, or the growing cycle for a particular crop. Harvesting is the most labour-intensive activity. Moreover, timely harvest before shattering of the produce or before rain assumes significance. As adjacent fields sown at the same time come to harvest simultaneously, farmers face great problem in getting manual labour for harvesting and cleaning their produce. To overcome these difficulties, different types of harvesters have been developed from time and again to suit different crops, situation and conditions.

6.2. Harvester and reaper

Harvesters are used to cut the crop after they mature. In such cases, threshing is to be done separately. Harvesters are available for a wide range of crops belonging to cereals, pulses, millets, oilseeds, fruits and plantation crops. Reapers are used to harvest crops such as rice, wheat, soybean, mustard, rapeseed, safflower and other thin stemmed crops. Reapers are operated using power tiller.

6.3. Combine harvester

The combine harvester, or simply combine, is a machine that harvests grain crops. It combines reaping, binding and threshing into a single operation. This was done individually earlier. In India, the crops harvested with a combine include rice and wheat. The waste straw left behind is either spread on the field or baled for feed and bedding for livestock.

6.4. Digger

It is used for digging tuber and root crops. It is efficiently used after cutting and removing the aerial stem portion of the plant. Various animal drawn, tractor-drawn and power-operated diggers are available.

6.5. Machineries and equipment used for harvesting

6.5.1. Manual

Berseem cutter (PAU), Citrus harvester (Push and cut type) (ICAR, Barapani), Citrus harvester (Hold and twist type) (ICAR, Barapani), Naveen sickle (CIAE), Tree climbing device (CPCRI), Ladder (To harvest arecanut -farmer's innovation), Improved shear (for tea harvesting)

6.5.2. Animal drawn

Groundnut diggers (CIAE, MPKV, TNAU), Birsa potato digger (BAU), Single-row potato digger (PAU), Multipurpose digger (CIAE), Reaper (CIAE)

6.5.3. Tractor drawn

Cotton stalk uprooter (PAU), Groundnut digger (APAU, CIAE), Groundnut digger shaker (PAU), Multipurpose vertical conveyor or reaper windrower (CIAE), Rear-mounted reaper windrower (CIAE), Self propelled vertical conveyor reaper windrower (walking type) (CIAE/TNAU), Combine harvester, Grain combine (PAU), Tractor mounted combine (Cruzer 7504-TAFE), Tapioca harvester (TNAU), Groundnut harvester (TNAU), Turmeric harvester (TNAU), Self propelled combine harvester, Soybean harvester (CIAE), Tractor operated combine harvester, Potato digger (TAFE), Two-row potato digger (PAU), Potato digger elevator (PAU), Orchard multi utility elevator platform (AAU), Tractor mounted telescopic hoist (AAU), Wheat straw combine (PAU), Stubble shaver (PAU)

6.5.4. Power operated

Multipurpose soybean reaper (CIAE), Riding type reaper (CIAE), Riding type small combine, Track-type small combine, Turmeric harvester (TNAU), Rice harvester (TNAU), Reaper (CIAE), Fodder sorghum harvester (TNAU)

6.5.5. Self-propelled

Austoft sugarcane chopper harvester, BUNMAI NB-11 TR sugarcane harvester, BUNMAI NB-11 BB sugarcane harvester, BUNMAI Sugarcane de-topper-cum-leaf stripper, CARIB sugarcane harvester, CAMECO S-30 sugarcane harvester,

CLAAS sugarcane harvester, CLAAS rice combine harvester, GK-75 whole stalk sugarcane harvester, Selective tea harvester (Williames Tea Pty Ltd)

6.6. Features of some commonly used harvesting machinery

Machinery	Features
Manual	
Berseem cutter (PAU)	Has a long handle made of hollow pipe and a curved blade fixed at right angles to each other. The width of the blade is 80 mm and the length is 650 mm. Single stroke covers an area of about 1.2 m wide and 0.6 m long.
Citrus harvester (Push and cut type) (ICAR, Barapani)	Consists of two blades, a bag for holding the fruits and blade actuating mechanism. The fruits can be harvested by shearing of the stem and carried through a holder bag to the ground.
Citrus harvester (Hold and twist type) (ICAR, Barapani)	The fruit is held in between the two halves of the harvester, which are actuated with the help of a rope and spring. Reduces the damage caused to the fruits while harvesting.
Coconut tree climbing device (CPCRI)	Used for climbing coconut trees to harvest the nuts. It consists of a wire rope passing through rubber pads. The climber can lock one of his legs at any height. The other leg is pulled up again and locked. It requires 1.5 minutes to climb a tree of 30 to 40 ft height.
Improved shear (Fig 6.1.)	The thickness of hinge points is more for both blades. This provides additional bearing /support area, minimises wear and tear rate. The dia of the pivot pin is 8 mm and made of gun metal. This reduces rusting and friction. Lock is provided near the pivot point which arrests free movement with respect to one blade, so that the relative motion & wear happens only on the pivot pin, with respect to the other blade.
Animal drawn	
Groundnut digger (CIAE)	Suitable for digging groundnut and potato. It is provided with two gauge wheels to control the depth of operation depending upon the crop being dug out. Various models of groundnut diggers designed by different institutes have a field capacity of 0.1-0.3 ha/day. It helps in saving 89% on time and labour and 71% on cost.
Birsa potato digger (BAU)	Suitable for digging potato tubers after removal of the aerial plant parts from the field. It consists of a ridge shaped bottom with welded extension rods on its wings. These rods separate soil adhering to the tubers. The handle at the rear guides the implement while in operation. It can penetrate to a depth of 16-18 cm. Damage to tubers while digging is only 3%. It has a field capacity of 0.03 ha/hr. Helps in saving 40% on time and labour and 18% on cost.

Multipurpose digger (CIAE)	Suitable for digging out groundnut and potato. The overall dimension (l x w x h) is 3600 x 850 x 850 mm. It has a field capacity of 0.05-0.12 ha/hr with 60% efficiency. It has a working depth of 20 cm.
Tractor drawn	
Rear-mounted reaper windrower (CIAE)	The reciprocating cutter bars harvest the crop which is collected on a tilting platform behind the cutter bar. The platform is released periodically by the tractor operator to enable the harvested crop to fall on the ground as heaps. The harvested crop is collected manually. Suitable for cereals, soybean and oilseeds. Overall dimension is 333 x 225 x 100 cm. Field capacity is 0.4 ha/h with an efficiency of 66%.
Grain combine (PAU)	Consists of a cutting unit, threshing unit, cleaning and grain handling unit. The crop after being cut is delivered to the cylinder and concave assembly through a feeder conveyor where it is threshed and grains and straw are separated into different sections. Suitable for harvesting wheat and paddy. Operated using 55 hp tractor or more. It has a cutter bar of width 3000-5000 mm and a grain tank of 800 kg capacity. It saves 80-90 % on labour and 33% on operating cost.
Tractor mounted combine (Cruzer 7504-TAFE) (Fig 6.2.)	Suitable for harvesting paddy and wheat with maximum recovery of full length straw. It has axial flow type of threshing system. Operated using 75 hp tractor or more with a 4-wheel drive. It has a cutter bar of width 3350 mm and a working capacity of 50 min/acre with a fuel consumption of 3 litres. No extra truck is required for transit.
Track type combine	This combine moves on rubber track instead of wheels and is suitable for harvesting in wet and soft land. Operated using 75 hp tractor. Ideal for harvesting paddy even when the field remains wet.
Wheat straw combine (PAU)	Used to recover wheat straw after combine operation. It is trailed behind the tractor and is PTO operated using a 35 hp tractor.
Groundnut harvester (TNAU) (Fig 6.3.)	Suitable for harvesting and windrowing of groundnut crop with 8-12% moisture content. It has a soil loosening attachment, pick conveying mechanism and gatherer windrower. The pickup conveying mechanism has endless ship chains spaced 1800 mm apart. At the rear, a gatherer windrower collects the conveying crop. It covers 2.0 ha/day with a harvesting efficiency of 99% and soil separation efficiency of 95%. The power source is a 35-45 hp tractor. It saves 32% on labour and 96% on time as compared to manual digging.
Turmeric harvester (TNAU)	Used for harvesting turmeric rhizomes. It has a five bar point blade with straight tynes at both ends to facilitate easy penetration into the soil. The rake angle of the blade is

	adjustable. At the rear end two converging slots are fixed to convey the harvested turmeric with the soil on to the lift rods without spilling sideways. After digging, the soil slips back to the ground and the dug out rhizomes are collected at the centre of the unit. It is operated with a 35-45 hp tractor. It covers 1.6 ha/day and causes minimum damage to the rhizomes (2.83%). It helps in saving 70% on cost and 90% on time as compared to manual digging.
Groundnut digger shaker (PAU)	Digs the groundnut plant below the pod zone and elevates them by an elevator-picker reel (conveyor) for dropping on the ground. The soil attached to the plant is shaken off in the process and a windrow is formed with the help of deflector rods. The plants are dropped in such a manner that the pods get exposed to the sun for speedy drying. The power source is from PTO of a 35 hp tractor or more.
Tapioca harvester (TNAU)	Before digging, the base is cut and the plant is removed. The depth of digging can be adjusted according to the tube geometry and moisture condition of the soil. Five pegs are provided at the front end of a trapezoidal digging blade for easy penetration into the soil and to reduce the draft requirement of the unit. Operated by a 35-45 hp tractor. Completes harvesting of 1.6 ha in a day.
Potato digger (TAFE) (Fig 6.4.)	Cut-opens the soil, lift the tubers and loads to a primary conveyor. The soil stuck to the tubers is loosened while passing over the rollers and drops down. The cleaned tuber is collected in bins. The collected tubers are periodically transferred to another gunny bag/container.
Power operated	
Rice harvester (TNAU)	Suitable for harvesting and windrowing of non-lodging rice varieties. Operated by a 3 hp kerosene engine. The overall dimension (l x w x h) is 2200 x 850 x 1170 mm. It has a cutting width of 750 mm and covers one ha in a day.
Riding type reaper (CIAE)	Suitable for harvesting rice, wheat, soybean and other crops. It consists of a crop row divider, star wheel cutter bar, conveyor belt and wire spring. Operated with a 6 hp diesel engine. The field capacity is 0.25- 0.3 ha/hr with 60-70% efficiency.
Multipurpose soybean reaper (CIAE)	Suitable for harvesting rice, wheat, soybean and other thin stemmed crops. Operated with a 8-10 hp power tiller. The cutter bar at the front of the reaper cuts the crop that is gathered to the loading platform by the standard reel. A flat belt moves over the platform to convey the crop to one side of the machine. It has a cutting width of 1200 mm. Field capacity is 0.09-0.12 ha/hr with 74% efficiency.

Self-propelled

Austoft sugarcane chopper harvester	Used for harvesting sugarcane grown at 5 feet spacing between rows. It has a field capacity of 0.24-0.30 ha/hr with a work efficiency of 39-44%. It gives an output of 24-30 t/ha.
CLAAS rice combine harvester	Used for both harvesting and threshing of paddy and wheat. It can also be used for crops such as soybean, gram, mustard and pulses. It has a 4.5 m wide cutting section and is equipped with crop lifters fitted at the knife section to ensure proper lifting of the plants and prevent damage to knife guards when operated on an uneven surface. It has a revolving tangential axial flow threshing and separation mechanism and a unique accelerator drum to optimize the crop flow from the feeder housing to the threshing area. It is fitted with a 125 hp turbo charged engine. It has 3000 litre grain tank capacity to store 1.8 t paddy or 2.7 t of wheat. The working width is 5060 mm.
Selective tea harvester (Williames Tea Pty Ltd) (Fig 6.5.)	Covers a width of 1500 mm in a single pass. Selectively plucks two mature leaves and a bud. Harvest efficiency is 4 km/h. Light weight and can be lifted across narrow drains. Collected leaves can be emptied in less than a minute. Operated by two persons. Fitted with light weight fuel efficient four stroke engine.

Fig 6.1. Improved shear

Fig 6.2. Tractor mounted combine – TAFE

Fig 6.3. Groundnut harvester – TNAU

Fig 6.4. Potato digger – TAFE

Fig 6.5. Selective tea harvester - Williames Tea Pty Ltd

7

Threshing

7.1. What?

Threshing is the process of separating grains from the straw or pod by impact, friction or both. Threshing should be done at an appropriate time after harvest. Delayed threshing causes rapid deterioration of the grains, especially during field drying or when the crop is stacked in the field. Improper threshing leads to wastage of grains and reduces the quantity of final marketable produce. After threshing, the produce has to be winnowed to remove lighter materials such as unfilled grains, chaff, weed seeds and straw using a blower, air fan or wind. Winnowing improves the ability of the grain to be safely stored, milling output and quality.

7.2. Methods employed

Traditionally, threshing is done by beating the harvested crop on a threshing floor directly or with a stick. Another traditional method of threshing is to make a bull walk in circles on the grain on a hard surface. In some areas, the harvested panicle with grain is spread on the surface of a country road so that the grain may be threshed by the wheels of passing vehicles. In these methods winnowing is separately performed. However, with the development of technology, now it is mostly done by machine referred to as threshers, shellers and stripper. Some of these threshers have inbuilt facility for threshing. Thus, threshing and winnowing are completed in a single process.

7.3. Machineries and equipment used for threshing

7.3.1. Manual

Castor sheller (APAU), Castor sheller (TNAU), Groundnut decorticator (CIAE),

Groundnut decorticator (TNAU), Octagonal maize sheller (CIAE), Tubular maize sheller (CIAE), Husking sheller (TNAU), Maize sheller (TNAU), Pedal operated thresher, Sunflower seed sheller (TNAU), Sunflower threshing bench (APAU), Phule sunflower thresher (MPKV), Sunflower thresher (UAS), Harambha thresher

7.3.2. Tractor operated

Groundnut thresher (PAU), Groundnut thresher (TNAU), Axial flow sunflower thresher (PAU)

7.3.3. Power operated

Power operated castor sheller (APAU), Castor sheller (GAU, Sardar Krishinagar), Castor decorticator (CIPHET), Groundnut decorticator (TNAU), Groundnut thresher (TNAU), Groundnut pod decorticator (CIPHET), Motorized rubber sheller (CIAE), Maize thresher (TNAU), Rotary maize cob sheller (CIPHET), Sunflower dehulling mill (CIPHET), Groundnut stripping bench (APAU), Comb type groundnut stripper (TNAU), Groundnut stripper (CIAE), Multicrop thresher (CIAE)

7.3.4. Power and tractor operated

High capacity multicrop thresher (CIAE), Multicrop thresher (APAU), Multicrop thresher (CIAE), Portable rice thresher (TNAU)

7.4. Features of some commonly used threshing equipment/ machineries

Threshers	Features
Manual	
Castor sheller (TNAU)	Consists of a feeding hopper, rubber-coated disc type shelling unit and a blower. It is suitable for shelling and winnowing of dried castor pods. The overall dimension (l x w x h) is 1320 x 1050 x 800 mm. It has a capacity to shell 165 kg castor in a day. Operated by a 0.5 hp electric motor.
Groundnut decorticator (CIAE)	Has an oscillating sector with three cast iron shoes and perforated screen. Decorticated pods fall through the screen and the kernels get separated manually. The decortication efficiency is 98-99%.
Octagonal maize sheller (CIAE)	Used for threshing maize. The sheller consists of 4 fins tapered along their length. The corners of the fins are rounded to avoid injury to the operator during shelling. The tapered edges of the fins dig into the space between the rows of the grains in

	the cob and, with the forward or backward stroke, the grains are released from the cob. One end of the sheller has a larger opening and the other a smaller one. It provides 100% shelling and cleaning.
Sunflower seed sheller (TNAU)	Suitable for both threshing and cleaning. Operated by a 3 hp electric motor. Blowing is effected by centrifugal blowers. The overall dimension (l x w x h) is 2.8 x 2.0 x 1 m. It has a shelling efficiency of 90% and separation efficiency of 96%. The output capacity is100 kg/hr.
Tractor operated	
Harambha thresher	A combination of beater type thresher and chaff cutter. This is popular in North India for threshing wheat. Chaff cutter blades cut the crop into pieces, while beaters help to detach the grain from the crop. Light materials like chopped straw are blown away with an aspirator blower, while the heavier materials such as grains and nodes fall on a set of reciprocating sieves. The sieves clean the grain and an auger elevates the grains and conveys directly into a trolley. Operated by a 35 hp tractor. It has a capacity to thresh 1.4 t/hr.
Groundnut thresher (PAU)	The unit has a threshing cylinder, centrifugal blower, cleaning shoe, set of sieves, conveyor and transport wheels. The dry crop is fed into the cylinder to separate the pods from plants. Pods and vines are moved into stammer saws of the cleaning shoe where the vine is cut into fine pieces. The pods are separated by screens and light material is blown away by an air blast. It is operated with a 35 hp tractor.
Power operated	
Castor decorticator (CIPHET)	Suitable for all varieties of castor. De-poding and decortications can be done simultaneously. There is zero damage to castor capsules.
Groundnut pod decorticator (CIPHET)	The unit consists of a feed hopper, decorticating drum, aspirator and sieve grader to grade kernel into two types. It has the capacity to decorticate and clean 60 kg pods/hr with an efficiency of 87% and no breakage of kernel.
Sunflower dehulling mill (CIPHET)	The unit consists of a centrifugal type dehuller, aspirator and grading unit. Dehulling, cleaning and grading are done in a single pass. It has a capacity to dehull 400-500 kg/hr.
Groundnut stripper (CIAE)	The stripper has a feeding tray, spike tooth threshing cylinder, concave grate, stripping bars, straw thrower, blower and collecting pan. Freshly harvested groundnut plants are fed in and the stripped pods along with light chaff and vines fall through the concave grate and treated by an air stream produced by a blower. The pods are collected in a pan and light materials are blown off.

Portable rice thresher (TNAU)	Used for threshing rice. Can also be used for threshing sorghum, bajra and safflower. The overall dimension (l x w x h) is 1.38 x 0.95 x 0.76 mm. It is operated using a 5 hp engine or electric motor.
Groundnut thresher (TNAU) (Fig 7.1)	Used to separate groundnut pods from the harvested plants. The thresher is an axial flow type and consists of a feed hopper, spike-tooth cylinder, concave, oscillating sieves and blower. The pegs are arranged on a cylinder in 10 rows. The oscillating sieves fitted below the cylinder separate the pods from the leaves, soil and other dust materials. The blower fitted in between the two sieves helps to blow out the leaves. It has a capacity to thresh 150 kg pods/hr with a threshing efficiency of 98% and cleaning efficiency of 92% . The damage to pods is only 1 to 3%. It saves 32% on cost and 70% on time as compared to the conventional method.
Power and tractor operated	
High capacity multicrop thresher (CIAE)	Suitable for threshing wheat, maize, soybean, sorghum, sunflower, pigeon pea, gram, mustard and other similar crops. The dry crop is fed continuously in bundles inside the thresher. Feed rollers provided in the feeding hopper push the crop into the threshing drum. Beaters or spikes on the threshing cylinder hit the crop and the impact causes detachment of grains from the ear heads of the crop. The aspirators blow out the chaff and lighter impurities through the outlet. Clean grain is obtained by passing through the sieves and secondary inlet of blower. It can be operated using a 20 hp electric motor or 35 hp tractor. It has a threshing efficiency of 99.3-99.8% and cleaning efficiency of 96.5-99.9%. Broken grain is restricted to 0.5-1.4%.

Fig 7.1. Groundnut thresher – TNAU

8

Postharvest Agro-processing

8.1. What?

Post harvest and agro-processing refers to activities carried out after harvesting a crop. These include milling, grading, processing and packing of the produce before it reaches the end users. This is also referred to as value addition, as it increases the marketable quality of the harvested produce.

8.2. Milling

It is the process of separating edible portion of the grain after removal of the husk or seed coat. This converts the products into a form that can be readily used for cooking, e.g. rice milling and dal milling.

8.3. Grading

Grading is the process of segregating the harvested produce based on the physical appearance including size and colour. Grading increases the market value of the crop.

8.4. Processing

Processing refers to conversion of primary agricultural products into other commodities for market. Surplus amount of the harvested produce can be processed and converted into a form that can be stored for a long time with minimum or no damage. For example, tomato can be processed as ketchup, oranges as squashes, and potato and tapioca as chips. In some cases, though the primary product can be stored for a long time, it is still processed and converted into some other form. For e.g. rice is processed into puffed rice, flattened rice, flour, etc.

8.5. Packing

Most of the processed commodities need to be packed in airtight pouches, free from moisture and insect pests, to increase the shelf life of the produce. Various kinds of packing materials such as polythene bags of different gauges, aluminium pouches, foils and pet bottles are used for packing.

8.6. Post harvest machineries used in various crops

8.6.1. Cereals

Parboiling unit (TNAU), Parboiling unit (CRRI), Improved parboiling unit (IICPT), Rice dehuller, Huller, Huller polisher, Husk stove (IICPT), Low friction huller (IICPT), Multipurpose yard drying implements (IICPT), Continuous rice puffing equipment (IICPT), Germ separator (IICPT), Pearling cones, Rubber roll sheller, Low cost extruder for snacks (CIPHET), Grain pearler (CIAE), Spectrum vibro cleaner, Pedal cum power operated grain cleaner (CIAE)

8.6.2. Pulses

Dal mill (CIAE, IIPR, TNAU), Multipurpose dal mill (CIAE), Dal mill cum wet grinder (TNAU), Dal scourer (Gota machine), Dal slitter, Mini dal mill, Mini grain mill (CIAE), Semi automatic mini dal mill, Water oil mixer, Soybean blanching unit (CIAE), Soybean flaking machine (CIAE), Manual Soybean dehuller (CIAE), Motorized Soybean dehuller (CIAE), Soy paneer pressing device, Soy milk powder protocol (CIPHET)

8.6.3. Oilseeds

Air screen groundnut pod cleaner, Groundnut pod grader (CIPHET), Pedal cum power operated grain cleaner (CIAE)

8.6.4. Sugarcane

Improved four roller sugarcane crusher (TNAU)

8.6.5. Tuber

Centrifugal granulator for cassava based feed (CTCRI), Cassava chopper (TNAU), Drying system using non conventional energy (CTCRI), Hand operated cassava chipping machine, Motorized tapioca chipper, Multi purpose mobile starch extraction plant, Pedal operated chipping machine, Cassava peeling knife (CTCRI), Pilot plant for liquid adhesive from tapioca starch, Expanding pitch rubber spool grader for potato, Potato peeler (CIAE), Potato slicer (CIAE), Potato pulper (CIAE)

8.6.6. Fruits

Fruit grader (CIAE), Aonla shredding cum stone extracting machine (AAU), Banana comb cutter (CIPHET), Banana fibre extractor (CTRI), Banana vacuum packaging (TNAU), Kinnow pilot plant for processing (CIPHET), Pomegranate aril extractor (CIPHET), Apple and mango grader (GBPUAT), Evaporative cooled storage structure (CIPHET), Harvester (IIHR), Raw mango slicer (IIHR), Raw mango cube cutter (IIHR), Raw mango peeler (IIHR), Raw mango grader (IIHR)

8.6.7. Vegetables

Tomato seed extractor, Brinjal seed extractor, Chilli seed extractor (CIPHET), Onion manual grader (NRCOG), Motorized grader (NRCOG), Tomato grader (CIPHET), Multipurpose axial flow vegetable seed extracting machine (PAU), Vegetable dryer (CIAE), Bhendi plucker

8.6.8. Plantation

Concentric type rotary grader RCN (NRC for Cashew), Steam boiler for RCN (NRC for Cashew), Coffee huller, Coffee pulper (TNAU), Manual pepper thresher (TNAU), Mechanical pepper thresher (TNAU), Coconut dehusker (CPCRI), Coconut dehusking machine (KAU), Coconut splitting machine (CPCRI), Coconut dehusker (farmer's innovation), Coconut splitter (farmer's innovation), Arecanut dehusker (CPCRI), Tender coconut punching machine (KAU), Snow ball tender nut machine (CPCRI), Bio mass fired dryer for coconut (CARI), Low cost solar dryer (CARI), Solar dryer (CPCRI), Small holder's dryer (CPCRI), Smoke free copra dryer (CPCRI), Electric dryer (CPCRI).

8.6.9. Multi purpose

Corn mill, Grain pearler, Grain mill (CIAE), Rotary sieve cleaner cum grader (TNAU), Sorghum peeler (TNAU), Pedal or power operated grain cleaner

8.6.10. Dryers

Batch dal dryer, Batch dryer, Multipurpose tray / LSU dryer (CIAE), Seed grain dryer (batch type), Multi purpose solar dryer (M S Frame) (CARI), Solar cabinet dryer (CIAE)

8.7. Features of some commonly used post harvest agro-processing machineries

Machineries	Features
Cereals	
Parboiling unit (CRRI)	Suitable for uniform parboiling of paddy grains. The overall dimension (w x h) is 760 x 110 mm. Grains are parboiled in batches @ 75 kg/batch.
Improved parboiling unit (IICPT)	Suitable for rural areas to produce minimally polished rice. It finds more use for parboiling organically produced rice. Both husk and firewood can be used as fuel source. It has a capacity of 50-60 kg/batch.
Rice dehuller	The rice dehuller is a composite machine which performs both dehulling and cleaning in single operation for which independent units are provided. The dehulled grains are automatically conveyed to the cleaning portion for removing the bran and other dust particles. Operated with a 3 hp motor. It can dehull 2-2.3 t paddy/hr.
Huller polisher	The rice huller polisher is similar to the rice huller, but has an inbuilt polishing unit. The polishing unit has emery rollers to polish the rice. Both hulling and polishing are done simultaneously. It is operated by a 16-18 hp motor. It provides an output of 225-300 kg cleaned rice/hr.
Rubber roll sheller	Used for shelling rice with high efficiency. It consists of rubber rollers driven by alloy steel gears. The output from the sheller does not get heated due to the provision of a double action cooling air blower cum exhaust system. This has been made mandatory by the Government of India. The capacity of the roll sheller is 1000 kg/hr. It is operated by a 2.2 kW power source.
Germ separator (IICPT)	Used for separating the germs from the brown rice. In each batch, 3 kg rice can be handled.
Continuous rice puffing equipment (IICPT)	Puffs the conditioned rice and produces silica-free, impurities-free expanded rice. The hot air produced is recycled, which helps in saving energy. IICPT is easy to operate and is pollution free. Produces 8 kg of puffed rice/hr.
Pulses	
Dal mill (CIAE) (Fig 8.1)	Used for de-husking and splitting of pigeon pea, black gram, green gram and lentil. The product to be milled is pre-conditioned by soaking in water and sun dried before it is fed into the mill. The grains fed into the hopper are milled in an emery/corundum roller and pass out through the outlet.

	Milling is completed in two passes. It has a milling efficiency of 88% and a capacity to mill 100 kg grains/hr. The percentage of broken grain is 3-5%.
Dal mill (IIPR)	The grains are soaked in water for 2-6 hours and sun dried for 1-2 days to attain a moisture content of 9-10%. These grains are then transferred to the hopper for dal making. The clearance between the discs is adjusted according to the type of grain, to obtain the optimum percent of de-husked split dal. The split dal, husk and powdered materials pass out of the machine from separate outlets. A cyclone is provided at the separator end for this purpose. Thus, split dal is obtained in a single pass. It has a capacity to split 75-80 kg grain/hr. It is used for making dal from various pulse grains such as pigeon pea, chickpea, black gram and green gram. The power required is 1.5 hp.
Mini dal mill	When pulses are fed between the two discs, they are split into two halves due to shearing action and collected through an outlet chute. It is used to split all kinds of legumes (soybean, pigeon pea, black gram and green gram) into dal. Its capacity is 30-40 kg/hr. Operated by a 0.5 hp motor.
Soybean blanching unit (CIAE)	A water heater with the provision for inserting and removing the perforated container is provided. Soybean is placed inside. A batch of 20 kg soy dal in four perforated cages is dipped into the annular space for 1 hr at 100°C. After treatment, the dal is air dried. This is used to impart wet heat treatment to soy dal and to eliminate anti-nutritional factors. Its capacity is 20 kg/hr.
Manual Soybean dehuller (CIAE)	The whole soybean is broken and split into dal. The split dal passes through the screen to remove the broken soy dal. The fan blows away the hulls and other light materials and separates the hull and broken pieces from soy splits. The hulling efficiency is 98% with a capacity of 35 kg/hr. Broken or damaged grains is 3-4%.
Soy paneer pressing device	Used for pressing and shaping the paneer. The device is a screw press type made of stainless steel and operated by turning a hand wheel provided on the top. The receptacle for paneer is perforated to drain out excess milk and water and give proper shape to the paneer. It has a capacity to press 16 kg paneer/hr.
Oilseeds	
Air screen groundnut pod cleaner	Used for cleaning groundnut pods by removing hollow pods, undersized pods and impurities from the raw mass. The capacity of cleaning is 1 t/hr. 2.3 kW power is required to operate the machinery.

Sugarcane

Improved four roller sugarcane crusher (TNAU)	Crushes sugarcane with an efficiency of 60-70%. It provides 8-10% more juice recovery as compared to conventional crushers. It is operated by a 7.5 hp electric motor. It has a capacity of 250 kg/hr.

Tuber

Cassava chopper (TNAU)	Suitable for cutting or slicing cassava into chips. It has a capacity to slice 270 kg/hr and is operated by a 0.5 hp electric motor.
Multi purpose mobile starch extraction plant	Consists of a hopper to feed the tubers, crushing disc or cylinder with nail punctured protrusions rotating inside the crushing chamber to crush the tubers, sieving tray to remove fibrous cellulose materials, stainless steel or plastic tanks to collect the sieved starch suspension and a tuber storage chamber. The addition of water during processing is controlled through a water pipe with holes fixed inside the hopper. It is operated by a 0.75 hp electric motor.
Potato peeler (CIAE)	Used for removing the outer skin of potatoes. Potatoes to be peeled are put inside the cylinder and rotated slowly. Due to the presence of rasps, the peels are removed. It can peel even ungraded potatoes. In 8-10 min one batch is completed. It has a peeling efficiency of 77% with a peeling loss of 4%.

Fruits

Fruit grader (CIAE)	Provided with an elevator feeder for constant feeding of fruits in grading mechanism. Fruits are separated into four grades based on their size. It is used to grade fruits such as apple, sweet lemon, oranges and large-sized sapota. It is operated by a 2.5 kW geared electric motor.
Aonla shredding cum stone extracting machine (AAU)	Used for extracting whole stone from mature aonla fruits and simultaneously obtaining aonla shreds. Aonla stones and shreds are conveyed forward along with the roller and are finally discharged separately at the other end. It has the capacity of shredding 60-70 kg aonla/hr. Recovery of shreds is 97-98% and that of stones is 93-94%. Operated by a 1 hp, 3 phase electric motor.
Banana fibre extractor (CTRI)	Extracts fibre from banana pseudo stem, leaf and peduncle. It consists of a rigid frame on which the rollers are placed. The rollers have a horizontal bar with blunt edges, which is connected to the motor by a belt pulley mechanism. It is operated by a single phase electric motor. It has a working capacity of 1.5- 2.0 kg/hr.
Pomegranate aril extractor (CIPHET) (Fig 8.2)	Whole fruits are fed into the extractor's inlet and the separated arils are collected at the outlet. It has a separation capacity of 500 kg/hr and a working efficiency of 90-94%.

	Damage to arils is only 1-2%. It is operated by a 0.75 kW motor.
Evaporative cooled storage structure (CIPHET)	Used for storing fruits and vegetables at a moderately low temperature and sufficiently high relative humidity for a short term. It uses wetted pads as the cooling medium and helps maintain 20°C lower temperature than the outside. It has negligible maintenance cost and consumes less electricity.
Raw mango cube cutter (IIHR)	Used for cutting raw mango into cubes weighing 6-16 g. It has a working capacity of 500 kg/hr with an efficiency of 87%
Vegetables	
Tomato seed extractor (CIPHET)	The fruits are crushed in the crushing assembly and passed on to a separator assembly, which separates the seeds and juice from skin. The seeds and juice are separated using a vibratory sieve mechanism and collected separately. Thus, the seed, juice and skin are obtained separately at three separate outlets. This can also be used to extract seeds from lime. 98% seed recovery is achieved. Its capacity is 45-60 kg/hr. Operated by a 1.5 hp, three-phase motor.
Motorized onion grader (NRCOG)	Consists of 2 sets of counter-rotating double rollers rotated by a 1 hp motor. It separates onions into 5 grades based on their size. The overall dimension (l x w x h) is 2100 x 1200 x 1800 mm. Its grading capacity is 1.5 - 2.0 t/hr.
Tomato grader (CIPHET)	The tomatoes when fed in roll down the pipes due to gravity and fall immediately wherever they find the right space according to their diameter. The collector is inclined at 10°, so that the tomatoes slide directly into crates. The important feature of a grader is its ability to adjust the gap between the pipes and the inclination of the grading table and hopper. It can also be used for other round fruits and vegetables. Tomatoes are graded into 4 sizes, *viz*., 25-40 mm, 40-55 mm, 55-70 mm and > 70 mm diameter. The overall grading efficiency is 66% with a capacity of 325 kg/hr.
Multipurpose axial flow vegetable seed extracting machine (PAU)	Used for extracting seeds from vegetables like brinjal, chilli, watermelon, summer squash and cucumber. The fruits are cut into small pieces in the primary chamber. These are further cut and crushed by axially arranged blades attached to a rotor shaft. The machine can extract seeds at a rate of 5.49, 3.78, 9.42, 4.68, 3.60, 6.60 and 1.42 kg/hr, respectively, for brinjal, tomato, chilli, summer squash, watermelon, squash melon and cucumber, respectively.
Vegetable dryer (CIAE)	The dryer is designed for drying high moisture crops such as cauliflower, cabbage and onion to achieve a low level of moisture content (from 90% to 6%). The overall dimension

is (l x w x h) 2720 x 965 x 2605 mm. Heat is generated by 16 heaters of 500 W and hot air is blown out through the blower operated by a 2 hp motor. It has 20 trays in which 50 kg vegetables can be placed at a time. Drying time varies from 11 to 14 hr.

Plantation

Concentric type cashew nut rotary grader (NRC Cashew)	Three sieve cylinders of different radii are used for grading raw cashewnut. Its operational capacity is 110 kg/hr with a grading efficiency of 93.20%.
Coffee huller	Used for hulling both parchment and dry cherry coffee. The coffee to be hulled is fed into the hulling cylinder where the cross beater forces the coffee to pass through the perforated screen. This results in the separation of the husk from the coffee beans. The coffee beans and husk then pass through a powerful aspirator, which can be adjusted precisely for the perfect separation of husk and peels from the coffee beans. The huller has a capacity of 1.0-8.0 t/hr. The power required to operate is 6.25 -17 hp.
Mechanical pepper thresher (TNAU)	Used to separate pepper berries from pepper vines. It has a threshing capacity of 320 kg/hr. It is operated by a 2 hp electric motor.
Coconut dehusker (CPCRI)	Provided with two sets of blades. Coconut is placed between these blades for de-husking. By operating the lever mechanism, the husk is split into 3 parts. It has a working efficiency of 100 nuts/man hr.
Arecanut dehusker (CPCRI)	The dehusker consists of a scissors mechanism which is operated with the leg while the nut is fed into the dehusker by hand. The working efficiency is 10 kg/hr for ripe nuts and 4 kg/hr for green nuts.
Bio mass fired dryer for coconut (CARI)	Consists of a combustion chamber, drying chamber and firing tray. Half split nuts are arranged facing up in a single row. Ignition is started at one end of the shell row. The coconut shell is used as a fuel source. In 18 hours, the moisture content can be brought down to 5.7% from 50%. In each batch, 1000 nuts can be dried.
Low cost coconut solar dryer (CARI)	This uses solar energy to dry coconuts. The nuts are kept in single row facing upward for drying. Drying is completed in 4 days and there will not be any fungal infection. This is useful for drying during all seasons. Drying can be effected at a temperature range of 36 to 52°C and solar radiation of 1173 w/m^2. In each batch 110 nuts can be dried.
Small holder's dryer (CPCRI)	This batch-type dryer uses indirect heat for drying coconut. Agricultural wastes such as coconut husk and shell are used as fuel. In each batch 400 nuts can be placed and drying completed in 36-38 hours.

Multi purpose	
Corn mill	Used for milling grains. The corn mill consists of a conical hopper through which the grain is fed into an oscillating chute operated by a cam. The clearance between the two discs is adjusted to control the fineness of the flour milled. A controlling screw with a check nut is provided on the operator's side for the purpose of adjusting the fineness of the flour. It can also be used for producing sooji. These are operated with an electric motor and flat belt pulleys. The speed of operation is 600 rpm and requires a power of 2.25 kW. The milling capacity is 250 kg/hr.
Grain mill (CIAE)	Used for grinding cereals and pulses and to produce grits/ flour. It is also used for obtaining coriander splits. It is operated by a power source of 1.0 kW.
Sorghum peeler (TNAU)	Used to remove the hull/seed coat from sorghum, ragi and other millets. It has an efficiency of 80-85% with a capacity of 25 kg/hr. It is operated by a 1 hp electric motor.
Pedal or power operated grain cleaner	Suitable for cleaning and grading wheat, sorghum, soybean, chickpea and pigeon pea. The purity of clean grains is 99%. Operated by a 0.5 hp single-phase motor. It can clean 500 kg of wheat, 900 kg soybean and 800 kg chickpea/hr.
Dryers	
Batch dal dryer	Used for drying various pulses such as pigeon pea, green gram, lentil and chickpea. The dryer consists of a drying unit and heating unit. The drying unit has twin-drying chambers with a partition, so that two varieties of dal can be loaded and dried simultaneously without any mixing. Trays in the drying unit have perforated floor through which hot air is distributed uniformly, ensuring even and thorough drying of dal. The drying temperature is 60-80°C. The capacity is 2-3 t/hr and moisture is reduced to 13%.
Multi purpose solar dryer (MS Frame) (CARI)	Suitable for drying coconut, black pepper, mushroom, green chillies, jack fruit bulb and fish. It saves 16% on time for black pepper, 31% for jack fruit bulbs and 37% for mushroom compared to conventional drying.

Fig 8.1. Dal mill – CIAE

Fig 8.2. Mechanical pomegranate aril extractor – CIPHET

9

Special Tools and Equipment for Horticultural Crops

9.1. What?

Most of the horticultural crops are raised by following special techniques such as budding and grafting. In case of landscape horticulture, frequent pruning and mowing are required to maintain the architecture. When new area is brought under cultivation, either jungle or unwanted plants have to be cleared. To perform all these operations, certain set of special tools are used. Some tools, though small, are of great need in raising and maintaining horticultural crops at various growth stages.

9.2. Special tools for horticultural crops

9.2.1. Manual

Axe, Dah, Hand shovel, Garden rake, Crow bar, Felling dao, Tea pruning dao, Billhook, Budding knife, Grafting knife, Budding and grafting knife, Pruning and slashing knives, Multipurpose chopping knife, Pruning secateurs, Pneumatic secateurs, Hedge shear, Hedge trimmer, Lopping shear, Forester shear, Grass shear, Garden sword, Flower scissors and coconut tree climber

9.2.2. Power/electric/tractor operated

Chain saw, String trimmer, Rotating disc mower, Cylindrical lawn mower (TNAU), Power tiller operated auger digger (TNAU), Post-hole digger (TAFE)

9.3. Features of some commonly used special tools

Tools	Features
Manual	
Dah	A hand tool used for cutting small trees and shrubs and clearing jungle-like growth. Dah is most commonly used in north-eastern hill states. It has a blade length of 360 mm and width of 60 mm at the handle and 30 mm near the edge. The thickness of the blade is 6.5 mm at the base and tapers to 3.15 mm at the cutting edge.
Billhook	The curved blade fitted to a wooden handle has single or double cutting edge. Cutting is accomplished through impact and shearing action. Used for lopping branches and cutting shrubs and other hard vegetative material.
Budding knife	The knife consists of a folding blade and a handle. The blade has two edges. One of the edges is sharpened all along its length, whereas the blunt or the other edge is sharpened at the tip and is slightly curved. This sharpened curved portion is used to create a 'T' opening or slot on the bark of the mother branch or twig for the insertion of the bud. The edge, sharpened all along its length, is used for cutting of scion stick or defoliating leaves from the scion and slashing of bud from the stick. The budding knives are available in various sizes according to the length of the blade.
Grafting knife	Resembling a household knife, it is mainly used to cut the scion sticks for veneer grafting, cleft and stone grafting and inarching. It can also be used to defoliate the leaves of the scion stick and make 'V' groove for grafting. Grafting knives are available in various sizes and are specified by the working length of the blade.
Budding and grafting knife (Fig 9.1.)	Used for both budding and grafting. It consists of two blades each for budding and grafting. The knife is also used for cutting thin unwanted twigs, defoliation of leaves and general cutting work in nurseries and orchards. The width of the blade is 15 mm for budding and 11 mm for grafting.
Pruning secateurs	Used to cut thick branches or twigs up to 20 mm diameter. The pruning secateur consists of two cutting blades or one cutting blade and an anvil, handle, volute spring to keep the blade and handle in open position and a locking device for keeping the secateur in closed position. There are a variety of secateurs such as single cut, double cut, parrot nose cut, roll cut, supa cut, replaceable blade type, easy cut and kiln cut. The secateurs are selected according to the operation and size of the twig or branch. They are available in various sizes, *viz.*, 150, 175, 200, 225 and 250 mm.
Pneumatic secateurs	These are pneumatically operated. Gripping blade of the shear is stationary and shearing action is imparted by the outer blade through the movement of piston, at the end of which is fixed a portable cylinder with high-pressure air. The device offers effortless, accurate and swift cutting. The double acting piston

	facilitates easy pruning of even large branches. It is used for pruning vines.
Hedge shear	Manually operated hand tool for pruning, trimming and cutting of hedges and shrubs. It consists of two blades. Some of the models are provided with pruning notch near the pivot of blades for cutting thick twigs. The hedge shear is used for pruning and trimming hedge and giving it a desired shape. Also used for cutting shrubs and removing haphazard growth in gardens and lawns. The blade thickness is 8 mm, while the length is selected according to the requirement.
Grass shear	The grass shear is a simple hand tool used for the maintenance of lawns and cutting soft vegetative material. The blades are joined to a 'U' shaped spring steel handle which keeps the shearing blades always in an open position. Cutting takes place due to shearing action of the blades. The overall length of the shear is 300 mm and the blade thickness is 2-3 mm.
Garden sword	The garden sword is a manually operated hand tool used in the sitting or squatting position. The sword consists of a metal strip or blade with one of the edges sharpened for cutting grass. Some of the garden swords are bent and thus can be operated in standing position. The garden sword is used for cutting grass in lawns. It can also be used for cutting or clearing bushes with thin and soft stems.
Power/electric/tractor operated	
Intercultivator	Suited for field preparation and intercultivation in hilly areas and has a low centre of gravity for ergonomic working posture. The intercultivator folds up easily for storage and transportation. The gearbox is protected from the ingress of dirt by a labyrinth seal.
Earth augur	Ideal for making pits to plant saplings in orchards and erection of fencing communication posts. It drills hole of 9" diameter and 2' depth. Its operation is faster, safer and most economical. Life of augur is 1800 pits in sandy soil and 800 pits in clay. It has very low weight and is easy to operate. Fuel consumption is 80 pits/ lit.
Bush/ Weed Cutter	Ideal for cutting dense undergrowth and unwanted weeds. It is suitable for tough and rugged terrains and allows to work through a full 360 degrees. It is light weight with less noise and low vibration. It has full harness for greater safety and less fatigue.
Rotating disc mower	Has a knife edge moving at a high speed which cuts the grass through impact. The cutting unit has two to four triangular or rectangular blades. The surface finish obtained with a rotating disc mower is rough as compared to that obtained with a cylinder mower. Therefore, this is normally used for rough areas such as orchards where the finish is not critical. Rotating disc mowers are available with electrical motors or driven by engine. It provides a cutting width of 340-380 mm and height of 30 mm.
Cylindrical lawn mower	Used for lawn maintenance. Consists of a cylindrical reel on the surface of which cutting blades are mounted in a spiral fashion. The cutting blades rotate and because of their spiral mounting cause progressive cutting action across the anvil blade. The grass is

	trapped between the rotating and anvil blades and cutting takes place due to shearing action. The machine has a front roller for adjustment of height of cutting and a grass box at the rear to collect the cut grass. It has a high-impact plastic wheel with rubber tyres.
Chain saw	This is a lightweight portable machine, also called power saw. Cutting is done by an endless chain fitted with cutters, which runs around a flat piece called the bar. The power is transmitted through a centrifugal clutch mounted on the crankshaft of the engine. The chain is of roller type and has left and right hand cutters spaced alternately along its length. In front of each of the cutters is a small projection called depth gauge which controls the depth of cut made by the cutter. It is used to trim dead or diseased wood from trees, remove unwanted, criss-cross branches or fell trees. It can be used for cutting branches of length up to 400 mm. Operated by 1.6 kW motor.
Telescopic tree pruner	Suitable for pruning and cutting of branches up to a height of *Ca.* 5 meter (17 feet) off the ground. Cuts up to 1 foot thick branches. It has an anti-vibration system for greater operator comfort and safety, a shoulder strap or full harness for greater safety and less fatigue and a telescopic shaft that can be compressed to convenient length for transportation.
Hedge trimmer	Ideal for trimming (cutting, pruning) trees, hedges or solitary shrubs (bushes) in hilly areas and gardens. It is suited for heavy pruning of hedges with thick stems and branches. Double-sided blades can be used for proper shaping. The hedge trimmer has low weight, very low vibrations and excellent balance for greater convenience. Fuel consumption: 1 lit/hr.
Mist blower	Suited for various plant protection functions for field, fruit and plantation crops, *viz*., liquid mist, dust blowing, wide dusting and ultralight volume spraying. It has a lightweight and compact construction and a hip belt for greater operator comfort. Fuel consumption: 4 acre/lit

Fig 9.1. Different types of budding and grafting knives

10

Miscellaneous Equipment and Machineries

This chapter deals with equipment and machineries that are used in certain specific operations/ areas.

10.1. Lac cultivation

High-capacity scraper cum grader, scraper cum grader

10.2. Mushroom production

Grain Cleaner, Grain boiler, Boiled grain and Chalk powder mixer, Bag/bottle filler, Spawn inoculators, Substrate mixer - manually operated

10.3. Fish

Mobile cool chamber (CIPHET)

10.4. Value addition

Snow ball tender nut machine (CPCRI), Coir briquetting machine, Coir fibre spinning machine

10.5. Features

Tools/equipment	Features
Lac cultivation (CIPHET)	
Scraper cum grader	Scrapes 3-4 lac sticks at a time and simultaneously crushes and separates the scraped lac into three grades. This does not give room for mixing of sticks and wood particles with the scraped lac. It is suitable for scraping non-uniform

	lac encrustations. Operated by a 1.5 kW electric motor. The working efficiency is 20 kg/hr.
High-capacity scraper cum grader	Scrapes10 lac sticks at a time and simultaneously crushes and grades the scraped lac into three grades. It is suited for scraping non-uniform lac encrustations. Operated by a 3 kW electric motor. The working efficiency is 50 kg/hr.
Mushroom production (NRCM)	
Grain cleaner	Has a capacity to clean 75-100 kg of grain/hr with an efficiency of 96%.
Grain boiler	Has a stainless steel tilting tank provided with a stirrer to mix the grains for attaining even boiling. Drainage valve provided at the bottom removes the water after completion of boiling. The tilting tank facilitates easy unloading of grain into the mixer. Each batch holds 100 kg of grains. The power required is 6 kW. It saves 64% power and 41% time compared to pan-type heaters.
Boiled grain and chalk powder mixer	Uniform mixing of chalk powder (calcium carbonate) with grain enhances better mycelial growth. It is provided with tilting-type drum for easy removal of grains. Operated by 1.5 hp electric motor. In cases of power failure, it can also be operated manually. In each batch, 100 kg of grains can be evenly mixed with chalk powder within 20 minutes.
Bag/bottle filler	Has a movable hopper with a chain pulley mechanism and provision for easy grain filling in bag/ bottle. It also has a convenient and revolving operator's seat. In an hour 460 packets are filled, each with 250 g of grain.
Spawn inoculators	Manually operated with a capacity to inoculate 105 bags/ hr/worker.
Substrate mixer	Used for mixing compost that is used as a substrate for mushroom production. One tonne of compost heap will be turned in 10-15 min. It consumes 1.2 kwh/tonne. This method helps in reducing lump formation (65%) and release of ammonia gas (35%). Labour and time required is reduced to 50% compared to conventional manual mixing.
Fish	
Mobile cool chamber (CIPHET)	Used for storing fish for retail marketing. The holding capacity of an insulated box is 8 plastic crates of size 540 x 360 x 295 mm in two layers of four each. The storage capacity is 150 kg of fish with 80% filling of each plastic crate and 1:1 ratio of ice and fish. Such units are already in use at CIBA, Chennai; CIFT, Cochin; NBFGR, Lucknow; CIFA, Bhubaneswar; CIFRI, Kolkata; CIFRI, Guwahati and CIFE, Mumbai.

Value addition

Snow ball tender nut machine (CPCRI)	The machine makes a groove (ring) in the coconut shell without breaking it. This exposes the kernel. The ball of kernel is then scooped out with a pliable blade. This facilitates serving of tender coconut without shell. Eight-month-old tender coconuts plucked from the trees are amenable for scooping.

11

Water Lifting Devices

11.1. What?

These devices are used to lift water to a height that allows access to water. Lifting devices can be used to lift groundwater, rainwater stored in an underground reservoir and river water. For agricultural use, water is lifted either from a borewell or open well to irrigate the crops. According to power sources, lifting of water can be accomplished either manually or by animal or electricity sources of power.

11.2. Human-powered devices

Man has a limited physical power output, which may be in the range of 0.08 to 0.1 hp. This power can be used to lift water from shallow depths for irrigation. These include swing basket, counterpoise lift, don, Archimedean screw and paddle wheel. However, nowadays these are not commonly used.

11.3. Animal-powered devices

Animal power is abundantly available in India. They are used to lift water, besides being put to use for other field operations and processing works. A pair of bullocks may develop approximately 0.80 horsepower. They can lift water from a depth of 30m or more. Of course, the rate of discharge will go down with increase in the lift. These include rope and bucket lift, self emptying bucket, two bucket lift, Persian wheel and chain pump.

11.4. Mechanically powered devices

Mechanically powered water lifting devices are usually termed as pumps, which are operated with the help of auxiliary power sources such as engine or electric motor. These pumps are capable of lifting large quantities of water to higher heads and are usually employed for irrigating horticultural crops. Basically, four principles are involved in pumping water: (1) atmospheric pressure (2) centrifugal force (3) positive displacement and (4) movement of column of fluid caused by difference in specific gravity. Pumps are usually classified on the basis of operation, which may employ one or more of the above principles. These include reciprocating pump, monoblock pump, end suction centrifugal pump, self-priming centrifugal pump, turbine pump, submersible pump, jet pump, sprinkler and drip irrigation systems.

11.5. Operational procedure of various water lifting devices

Water lifting device	Operation
Animal powered	
Rope and bucket lift	Used to lift water from lined wells up to a depth of 30m. The device consists of a bucket or bag made of GI sheet or leather, and a pulley arrangement. A rope is attached to the bucket or bag, which passes over a pulley and finally fixed to the yoke of bullocks. The bullocks walk down on an earthen ramp sloped at an angle of 5-10° to lift water. About 9000 1/hr can be lifted with two pairs of bullocks from a depth of 15 m. The lift is also known as Mothe, Charsa or Pur.
Two bucket lift	Two buckets are raised and lowered alternately. The bullocks move in a circular path and with the help of a central rotating lever, rope and pulley arrangement the buckets move up and down. Each bucket may have a carrying capacity of up to 70 litres. Guide rods are provided in the well for the movement of buckets. The buckets are automatically filled and emptied during operation. The device is suitable for lift up to 5 m, at which discharge will be up to 14000 l/hr.
Persian wheel	Used to lift water from a depth of up to 20m. It has an endless chain of buckets made of GI sheet, each with a capacity of 8-15 litres. The chain of bucket is mounted on a drum and submerged in water to a sufficient depth. The animals move in circular fashion rotating the buckets carrying water through the gear system. The water is released when the bucket reaches the top. The average discharge of a Persian wheel is about 10,000 l/hr from a depth of 9 m with a pair of bullocks.
Mechanically powered	
Reciprocating pump	Used for both irrigation and drinking water supply. The reciprocating pumps are available in various designs and models, and can be operated manually, with animal power and auxiliary power sources. These are used to lift water from underground sources; therefore, if the water level is deep, the pump is to be lowered close to the water surface to reduce the suction head.

The number of cylinders can be increased according to power sources. Though it can be used for lifting water from 62m depth, the optimal depth is 28-34 m. It discharges 12-14 l of water per minute.

Monoblock pump

It is one of the most common types of centrifugal pumps employed for irrigation. It consists of an impeller or rotor and a progressively widening spiral or volute casing. The pump is directly connected to the prime mover, which may be an electric motor or engine. Upon rotation of the impeller, the water enters at the eye and is thrown radially outward to the periphery. Such an action causes vacuum at the eye and thus more water enters the suction pipe to maintain a continuous flow. Operated by a 1.5-7.5 kW motor. It discharges 2-7 litre per second.

Self-priming centrifugal pump

These are available in stationary and portable models. Primarily used where low-discharge is needed at higher heads. Due to self-priming features, the pump can handle air and gases entrained in water. These pumps are designed for handling water and do not require any foot valve. They have high suction lift characteristics and can suck water from a much lower depth. The pump can be coupled with an electric motor or engine. The pump is used for water supply in hilly areas and in sprinkler irrigation systems. The stationary type is operated using a 0.37-22.0 kW power source with a discharge rate of 10-600 l/min, while the portable type is operated using a 1.5-5 hp engine with a discharge rate of 380-1115 l/min.

Turbine pump

These pumps are used in tube wells or in open wells where the water level is deep. The pump unit remains submerged in water, whereas the prime mover (motor or engine) is kept above the ground level. The pump unit is connected to the motor with the help of a long vertical shaft supported on bearings, which may be water or oil lubricated. The gradual enlarging vanes guide the water to the casing, thus converting kinetic energy into potential energy. Therefore, the turbine pump generates heads several times that of a volute pump. Turbine pumps are most effective for tube wells and applications requiring high heads and discharge. It discharges 150-62000 l/min.

Submersible pump

These are also turbine pumps in which the long vertical shaft connecting the motor and pump unit is replaced by a short shaft. The prime mover and pump become closely coupled and submerged in water. Submersible pumps are suitable for tube wells having a bore of 100 mm or more. Submersible pumps consume less power for the same output and also require less space. Varying discharge rates can be achieved based on the pump size and the number of stages.

Jet pump

This is a diffuser pump which is used to lift water from both shallow and deep wells. During working, the output of the diffuser is split and half to three-fourth of the water is sent back down the well through the pressure pipe. At the end of the pressure pipe, water is accelerated through a cone-shaped nozzle. The water goes through a venturi in the suction pipe. The venturi speeds up the water causing a pressure drop, which sucks in more water through the intake at the very base of the unit. It uses a 0.5-1 hp motor and discharges 14-45 l/min.

Sprinkler irrigation system

Irrigation by sprinkler is nearest to natural rainfall, where water is sprayed into air in the form of coarse droplets. Water is subjected to pressure by pumping and discharged through small orifices called nozzles, which break the liquid into coarse droplets. Nozzles are mounted on a rotating head. It should not be used for fine textured soil where the infiltration rate is less than 4 mm/hr. Soluble fertilizers, herbicides and fungicides can also be applied using the sprinkler system. The system is also suitable for orchards, tea and coffee plantations. Power drawn is usually high as the sprinkler head operates under 0.5-10 kg/cm^2 pressure. Irrigation efficiency is 80%. The portable system is more popular because it can be moved after irrigating one crop to the other and also requires less initial investment. The perforated system is suitable for lawns, vegetables fields, gardens and crops where the plant height is below 60 cm. It requires clean water to avoid clogging of holes. The spacing of the sprinkler head on the lateral lines is decided based on the average wind speed of a particular area. The sprinkler system cannot be efficiently adopted where the wind velocity is very high.

Drip irrigation system

Drip irrigation is suitable for orchard, vegetable and other crops grown in rows. More than 90% of the water applied is efficiently utilized by the plants. Water is applied in the form of drops or very fine stream, thereby limiting the discharge to 2 to10 l/hr to a single plant. The application is almost equivalent to the consumptive use of plants. 20 to 30% more yield can be achieved with this system and 40 to 60% more area can be irrigated with the same quantity of water as compared to surface irrigation. Soluble fertilizer and pesticides can also be applied along with the water. The water can be applied on the surface or subsurface, very near the root zone of the plant. The pipelines are made of black PVC to avoid growth of algae in the lines. The water should be clean and free from any debris to avoid clogging of pipes, emitters or micro tubes. The laterals to which emitters or micro tubes are attached may have a pressure as low as 0.15 to 0.2 kg/cm^2 and as high as 1 to 1.75 kg/cm^2.

Others

Hydraulic ram

The hydraulic ram is a device that lifts water without any prime mover by utilizing the kinetic energy of flowing water. In this system the impact of water is converted into shock waves, which is called water hammer. This energy is utilized for lifting water.

11.6. Points to be considered before installation and selection of pumps

- Source of water supply: open well, tube well, canal, etc.
- Depth of water table in the area
- Crops grown
- Total cultivated area
- Type of prime mover (engine or motor). In case of electric motor, the hours of electric supply.
- Location of tube well
- Type of drive (belt driven, direct coupled, monoblock)
- Water conveyance system: lined or unlined or underground
- Groundwater quality

11.7. How to select the pump?

- Estimate the depth of water source (borewell/open well/canal, etc.) to be pumped and the delivery head. Know the discharge rate in case of borewell/open well.
- Note the specifications of manufacturers.
- Determine the type and capacity of pump based on suction head, discharge and delivery expected from the water source. Pumps have efficiencies ranging from 50 to 70%.
- Go for ISI specifications

11.8. Guidelines for efficient use of pumps

- Install the centrifugal pump 1-2 m above water level in the tube well
- Use large radius bends
- Do servicing and annual maintenance of pump set as per manufacturer's instructions
- To avoid leakage in joints, tighten the joints properly; use good quality gaskets
- Run the pump with the recommended number of revolutions
- Use good-quality driving belts, if it is a belt-driven pump

- Use proper size of suction and delivery pipes according to discharge
- Use good quality reflex valve whose flap should open fully
- Foundation should be strong, levelled and done well with bolts embedded in it
- Align the motor and pump pulley accurately.

12

Calibration of Seed Drills

12.1. What?

Seed drills are designed to sow seeds at the recommended seed rate with precision and at the desired depth. Proper adjustments are made to the seed metering system to ensure efficient use of seed drills. This aspect is vital to have a good crop stand.

The seed size varies between crop and their varieties. In addition, the seed metering systems are based on volume displacement. Therefore, if one lot of seeds varies in size and weight from another, two different amounts or number of seeds will be metered if the drill setting is not changed. For this reason, metering systems should be calibrated each time when put to use.

The method of calibration will be given in detail in the manual provided along with the seed drill.

12.2. Calibration of seed drill

- Measure the width of the seed drill in feet
- Decide on the seed rate to be used per acre
- Set the seed drill to the desired seeding rate
- Collect the actual seeding rate of the drill

Run the tractor attached with the seed drill for a distance of 100 ft. While moving collect the seed in a bag attached to the seed drop point. Weigh the quantity of seed collected in this run.

- Determine the strip length

Strip length (in field) = Distance driven to collect seed, this is usually 100 ft

- Determine the current seeding rate of the seed drill using the formula

$$\text{Seeding rate (kg/ac)} = \frac{43560\ \text{ft}^2/\text{ac x kg of seed collected}}{\text{Width of drill (ft) x Strip length (ft)}}$$

When both sides of the equation are equal, then the seed drill is calibrated. If the numbers on either side of the equation are not equal, the drill needs to be adjusted and the process is repeated until the drill is calibrated (Chapman and Carter, 1976)

Example

A seed mixture needs to be planted at a rate of 15 kg per acre. A tractor with a 15 ft seed drill was driven 100 feet and 0.4 kg of seed was collected.

Required seed rate = 15 kg/ac

Width of seed drill = 15 feet

Strip length = 100 feet

Quantity of seed collected = 0.517 kg

Current seeding rate =

$$15\ \text{kg/ac} = \frac{43560 \times 0.517}{15 \text{ X } 100}$$

$$15\ \text{kg/ac} = \frac{17424}{1500}$$

15 kg/ac = 15.01 kg/ac

As both the sides of the equation matches, the seed drill is calibrated properly to deliver 15 kg/ac.

If the equation was not equal, and the drill was delivering either more or less quantity of seed than the required seed rate, then, the seed drill needs to be set at a higher rate or lower rate, accordingly and recalibrated until the equation is equal.

12.3. Validation of the calibration

- Fill the seed box with seed and set the indicator at the desired seed rate according to the chart given by the manufacturer.
- Mark a distance of 10 to 20 meters in the field.

- Run the seed drill and collect the seeds in each tube for 10 to 20 meter length of run.
- The amount of seed collected in each tube in 10 or 20 meter run is then expressed in g/meter.
- This quantity should be equal to the calculated seed quantity obtained from the nomograph in g/meter length/row (Fig 12.1.).
- If the measured quantity is less or more, adjust the rate with the help of a seed metering lever.
- Sliding the roller out will increase the seed rate.
- Now, the seed drill is ready for planting the specific seed.

For this, it is better to calibrate the number of seeds with the number of revolutions of the drive wheel. If the circumference of the drive wheel is 1 m, then the length covered in one turn is also 1 meter and x number of seeds will fall.

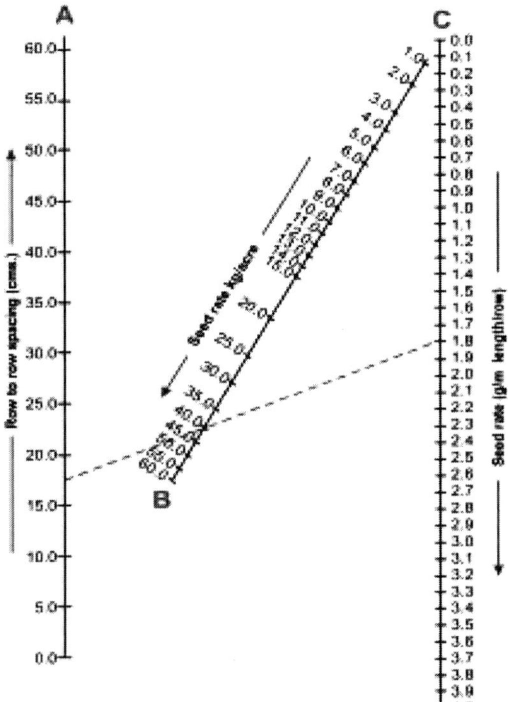

Fig 12.1. Nomograph for seed drill calibration (Extracted from Yadav *et al.*, 2002)

13

Maintenance

This chapter deals with the maintenance, repair, safety and trouble shooting of agricultural implements, equipment and machinery.

Routine maintenance reduces repair bills and extends the life of equipment. Good maintenance practices are essential for efficient operation of all types of farm machineries. Effort spent in this area of farm management is more than repaid by the consistent, reliable operation of machinery, reduced fuel bills and extended life of equipment. Maintenance of farm machinery is complicated by the usage pattern of short spells of intense activity, followed by periods of non-use or storage.

During the "standing" or non-use periods chemical interactions between metals and fluids can cause more damage than normal wear and tear from active usage. This must be considered in planning machinery maintenance.

- Inspect machinery at the end of season or harvest.
- Repair and adjust as required.
- Carry out maintenance work without pressure between seasons.
- Oil as and when required.
- Clean implements at weekly intervals.
- Change shovels and blades as soon as it is worn out.
- Do not work with implements having worn out parts.
- Do not use implements on hard surface.
- Do not tow the implement on road.
- Do not allow rusting.

For sprayers

- After completion of spraying, wash the sprayer and nozzle well with water.
- Ensure that all the settled pesticides are washed out properly.
- Discard all waste washings properly.
- If spray fluid is spilled on the sprayer during loading or mixing, wash the outside of the sprayer immediately.
- As a general rule, plastic or polyethylene tanks and hoses require extensive cleaning than stainless steel tanks.
- Screens and strainers should also be cleaned or replaced as and when required.
- Residues can also accumulate in cracked hoses. Inspect the inside of hoses and replace if necessary.
- Thorough cleaning (even with soap solution) is essential when there is switchover from one crop to another or from pesticides to herbicides and vice versa.
- If clogging is observed while spraying, check for any blockage in the nozzle.
- Always use washers to prevent leakage from hose and nozzle. If worn out, replace immediately.
- Apply oil to the trigger cutoff valve and priming lever for smooth operation.

14

Appendices

Appendix – I

Glossary of terminologies related to tractor

Term	Definition
3-point linkage	A standardized mechanism for attaching implements, consisting of an adjustable top (pivot) link centred above two lower (lifting) links.
Ampere	A unit of measurement used in expressing the rate of electrical current flow in a circuit. It is determined by dividing the voltage by the resistance.
Articulated body	A tractor body that flexes in the middle to provide a tighter turning radius
Baffle	An obstruction (e.g. plate or vane) used to slow down or divert the flow of gases, liquids, sounds, etc. It is found in the fuel tank, crankcase, muffler and radiator.
Ballast	Weights added to the tractor or implement to improve balance, traction, stability or digging force.
Bearing	The area of a unit in which the contacting surface of a revolving part rests to minimize wear and friction between two surfaces.
Belt horsepower	The power of the engine measured at the end of a suitable belt pulley or power take-off shaft, receiving drive from the PTO shaft of the tractor
Bevel gear	A gear shaped like the wide end (frustum) of a cone, used to transmit motion through an angle. It is found in differentials.

Bore	The diameter of the cylinder.
Bore-stroke ratio	The relation between the diameter of the cylinder bore and the length of the stroke of the piston
Brake	The mechanism that converts motion (kinetic energy) into heat energy through friction.
Brake assembly	An assembly of the non-rotational components of a brake including its mechanism for development of friction forces.
Brake drum	A cast iron or aluminium housing bolted to the wheel that rotates around the brake shoes.
Brake horsepower	(bhp) A measurement of the actual usable power (not calculated power) measured at the output shaft (usually the crankshaft) rather than at the driveshaft or the wheels. Thus, none of the auxiliaries (gearbox, generator, alternator, differential, water pump, etc.) are attached. It is called the brake horsepower because the shaft power is usually measured by an absorption dynamometer or brake. This is not the brake on the vehicle's wheels, but a testing device applied to the shaft. This instrument is applied to stop or absorb the rotation of the output shaft and returns a value. One bhp = 746W or (approximately) 3 bhp = 4kW.
Brake lining	A heat-resistant friction material (usually asbestos) that is attached to the brake shoe (either riveted or bonded). When the shoe is pressed against the brake drum, the lining grabs the inside of the drum, which stops the vehicle and also prevents the drum and the shoe from wearing each other away.
Brake shoe	That part of the brake system, located at the wheels, upon which the brake lining is attached.
Cab-over Engine	A truck or tractor design in which the cab sits over the engine on the chassis. The cab-over is identified by the windshield being located directly over the front bumper and the driver's seat is directly over the steering axle.
Calcium	Short form of calcium chloride–water solution used to fill tractor tyres for ballast.
Calibration	Marking the measuring units on an instrument or checking their accuracy.

Caliper	The clamping device on disc brakes which straddles the rotating disc and by hydraulic action presses the pads against the disc to stop or slow the vehicle.
Cam	A metal disc with irregularly shaped lobes used to activate the opening and closing of the valves.
Camshaft	A shaft with cam lobes (bumps) which is driven by gears, a belt or a chain from the crankshaft. The lobes push on the valve lifters to cause the valves to open and close. The camshaft turns at half the speed of the crankshaft.
Camshaft drive	A connection between the crankshaft and camshaft by means of gears, chain, drive belt, shaft or eccentric shaft to maintain the ratio of 12.
Camshaft housing	That part of the engine which encloses the camshaft and often other parts of the valve train.
Carburetter	A device that vaporizes fuel and mixes it with air in proper quantities and proportions to suit the varying needs of the engine.
Clutch	A device that disconnects the engine from the transmission, to allow the vehicle to change gears and then allows the engine and transmission to resume contact and turn together at a new speed.
Clutch housing	A cast iron or aluminium housing that surrounds the flywheel and clutch mechanism.
Clutch plate	A spinning plate located at the end of the driveshaft facing the engine flywheel and covered with a friction material such as asbestos.
Crank	An arm set at right angles to a shaft or axle, used for converting reciprocal (to-and-fro) motion into circular motion.
Crankshaft	A main rotating shaft running the length of the engine. The crankshaft is supported by main bearings. Portions of the shaft are offset to form throws to which the connecting rods are attached. As the pistons move up and down, the connecting rods move the crankshaft around. The turning motion of the crankshaft is transmitted to the transmission and eventually to the driving wheels.

Crankshaft gear	A gear mounted on the front of the crankshaft. It is used to drive the camshaft gear.
Crown wheel and pinion	A pair of gears in the final drive of a vehicle, always found in the back axle of a rear-wheel drive layout where the pinion is on the end of the propeller shaft driving the crown wheel mounted on the differential at right angles to it, and also in front-wheel drives where the engine is not transversely mounted.
Differential	A unit on rear-wheel drives vehicles that takes the power of the rotating driveshaft at right angles to the rear axle and passes it to the axle. It will not only drive both rear axles at the same time, but will also allow them to turn at different speeds when negotiating turns. In this way the tyres do not scuff or skid
Differential lock	A device for temporarily locking the rear differential to achieve better traction.
Dipstick	The metal rod that passes into the oil sump. It is used to determine the quantity of oil in the engine. The oil level is marked on the rod and matches level indicators on the rod. Dipsticks are used to check engine oil and transmission fluid.
Disc brake	A type of brake that has two basic components: a flat rotor (disc) that turns with the wheel and a caliper that is stationary. When the brake pedal is depressed, linkage (mechanical or hydraulic) causes the caliper to force its heat-resistant brake pads against both sides of the rotating disc, thus slowing or stopping the wheel.
Draft control	A mechanical linkage that maintains an implement at a constant working depth in spite of varying working conditions.
Drain plug	Usually, a threaded plug at the lowest point of the sump, gearbox, cooling system, etc., which is removed to drain the oil or coolant and typically has a recessed hexagon head.
Drawbar	A flat horizontal bar attached to the rear of the tractor, used for dragging or towing.

Drawbar horsepower	It is the horsepower measured at the end of the tractor's drawbar. This is the power available to pull loads.
Drive gear	The gear which transmits the power to a driven gear.
Driven gear	An engine needs to transmit power to the wheels by the use of sprockets and chain (as in a motorcycle) or by a drive gear which meshes with a driven gear to propel the vehicle.
Drum brake	A type of brake using a shallow drum-shaped metal cylinder that attaches to the inner surface of the wheel and rotates with it. When you press down on the brake pedal, curved brake shoes with friction linings press against the inner circumference of the drum to slow or stop the vehicle.
Dual brakes	A brake system that uses a tandem or dual master cylinder to provide a separate brake system for both front and rear of vehicle.
Engine	A machine for changing fuel energy to mechanical energy.
Engine coolant	Antifreeze liquid used in the engine's cooling system
EROPS	Enclosed ROPS. Air-conditioned cab or heated cab with ROPS.
Fender	A covering over the wheels to prevent mud from splattering.
Flywheel	A relatively large and heavy metal wheel that is attached to the back of the crankshaft to smoothen out the firing impulses. It provides inertia to keep the crankshaft turning smoothly during periods when no power is applied. It also forms a base for the starter ring gear and, in manual transmission, for the clutch assembly.
FOPS	Falling Objects Protective Structure. A heavy duty structure for protection of the machine operator or driver from falling objects. Usually has four posts and a strong roof
Friction	Resistance to motion between two bodies in contact with each other.
Friction horsepower	(FHP) The amount of power consumed by an engine in driving itself. It includes the power absorbed in mechanical

	friction and in driving auxiliaries plus, in the case of four stroke engines, some pumping power.
Governor	A governor is essentially a speed-sensitive device, designed to maintain a constant engine speed regardless of load variation, by controlling the rate of fuel delivery. It is also known as a speed limiter.
Helical gear	A gear that has the teeth cut at an angle to the centre line of the gear. This kind of gear is useful because there is no chance of intermittent tooth-to-tooth operation as at least two teeth are engaged at any time. Also, helical gears tend to operate quieter than straight-cut gears.
Hitch	The bracket used to connect a vehicle to a trailer.
Horsepower	(HP) A measurement of the engine's ability to perform work. One horsepower is defined as the ability to lift 33,000 pounds by one foot in one minute. 1 HP is also equal to 745.7 watts.
Hydraulics	A system of pressurized oil which provides power for raising and lowering the three-point hitch and which can be used to operate attached or towed implements having hydraulic pistons and cylinders.
Hydrostatic transmission	A hydraulic transmission which varies the tractor's forward or reverse speed within a gear range based on pressure applied to a lever or pedal.
Ignition system	Electrical system devised to produce timed sparks from engine spark plugs, consisting of a battery, induction coil, capacitor, distributor, spark plugs and relevant switches and wiring.
Implement	A tool or device that is attached to or towed by a tractor.
Indicated horsepower (Ihp)	A measure of the power developed by the burning fuel within the cylinders. Ihp includes bhp plus the power lost to friction, and pumping needed for the induction of the fuel and air charge into the engine and the expulsion of combustion gases
ISOBUS	ISOBUS is a communication protocol for the agriculture industry. It specifies a serial data network for control and communications on forestry or agricultural tractors and

	implements. It consists of several parts: General standard for mobile data communication, physical layer, data link layer, network layer, network management, virtual terminal, implement messages applications layer, power train messages, tractor ECU, task controller, management information system data interchange, mobile data element dictionary and diagnostics
Loader	A bucket loader attachment, usually hydraulically operated and front mounted
Maximum horsepower	The horsepower measured at the engine flywheel without attachment of any of the power-consuming accessories.
MFWD	Mechanical Front Wheel Drive (i.e.) four-wheel drive.
Net horsepower	The horsepower measured at the engine flywheel with all the power-consuming accessories attachment to the tractor.
Nozzle	A mechanical device designed to control the direction or characteristics of a fluid flow as it exits (or enters) an enclosed chamber or pipe via an orifice.
Overrunning clutch	A safety mechanism which allows a heavy implement to spin down without energizing the tractor-driven train.
Power Take Off	(PTO) An extension of the drive train that allows power to be mechanically transferred to other machinery or implements via a removable driveshaft with splined couplings.
PTO horsepower	Indicates the actual usable power required by the tractor to operate implements.
Rear-axle	The shaft on which the back wheels revolve.
Remotes	Hydraulic inlet/outlet pairs at different locations on the tractor for connecting hydraulically operated implements.
Ripper	Large, heavy hooks behind the tractor for ripping roots or brush. Commonly used on bulldozers clearing land.
ROPS	Rollover Protective Structure: a roll bar or similar structure to protect the driver in case the machine tips over
Self-energizing brakes	Brakes which use the tractor's momentum to increase the braking force.

Sleeves	Tubular inserts placed into the engine block to serve as the cylinder bore. This allows complete replacement of the cylinder bore without re-boring the block to a larger size.
SMV emblem	The ubiquitous orange triangle symbol, which denotes a slow moving vehicle, required on the rear of farm machinery being driven on public roadways as a caution to traffic approaching from the rear.
Stroke	Displacement of piston from the top dead centre to the bottom dead centre.
Throttle	The mechanism by which the flow of a fluid is managed by constriction or obstruction.
Tractor Loader Backhoe (TLB)	A tractor, usually industrial, equipped with a front mounted loader and rear-mounted backhoe.
Torque	Force applied on a point to cause a turning effect. The unit of torque is one pound-foot.
Turbo intercooler	A radiator-type device for cooling air before it enters the engine. Cool air is denser allowing more oxygen density and therefore better performance.
Turbocharger lag	The time required to change power output in response to a throttle change
Turning radius	The size of the smallest circular turn a tractor can make.
Variable speed governor	A governor which is capable of holding any speed between idling and maximum speed.
Zerk fitting	A fitting or nipple provided for injecting grease into a mechanism.

Appendix-II

Addresses

This chapter deals with the addresses of ICAR institutes and State Agricultural Universities who have developed prototypes of various implements, equipment and machinery.

If someone is desirous of getting the implement, equipment or machinery fabricated, then, they have to approach the following institutes to get the addresses of commercial manufacturers identified by them at their respective places. Abbreviation of the institute which developed the products listed in this book is indicated within brackets.

ICAR institutes and State Agricultural Universities

1. Director, Central Agricultural Research Institute (CARI), P.B.NO.181, Port Blair - 744 101, Andaman and Nicobar Islands
2. Director, Central Institute of Agricultural Engineering (CIAE), Nabi Bagh, Berasia Road, Bhopal - 462 018, MP
3. Director, Central Institute of Post Harvest Engineering and Technology (CIPHET), Ludhiana 141 004, Punjab
4. Director, Central Tuber Crops Research Institute (CTCRI), Sreekariyam, Thiruvanthapuram 695 017, Kerala
5. Director, Directorate of Cashew Research (Earlier National Research Centre for Cashew), Indian Council of Agriculture Research, Darbe P.O., Puttur- 574 202, Dakshina Kannada, Karnataka
6. Director, Directorate of Mushroom Research (Earlier National Research Centre for Mushroom), Indian Council of Agriculture Research, Chambaghat 173213, Solan , Himachal Pradesh
7. Director, Directorate of Onion and Garlic Research (Earlier National Research Centre for Onion and Garlic), Indian Council of Agricultural Research, Rajgurunagar 410 505, Pune, Maharashtra
8. Director, Indian Institute of Crop Processing Technology (IICPT) (Ministry of Food Processing Industries, Govt. of India), Pudukkottai Road, Thanjavur 613 005, TN

9. Head, Agricultural Engineering Extension, College of Agricultural Engineering and Technology, Anand Agricultural University (AAU), Godhra 389 001, Gujarat
10. Head, Agricultural Engineering section, Indian Institute of Pulses Research (IIPR), Kanpur - 208024, UP
11. Head, Agricultural Machinery Research Centre, Tamil Nadu Agricultural University (TNAU), Coimbatore - 641 003, TN
12. Head, Department of Agricultural Engineering, Birsa Agricultural University (BAU), Kanke, Ranchi - 834 006, Jharkhand
13. Head, Department of Agricultural Engineering, Gandhi Krishi Vigyan Kendra, University of Agricultural Sciences (UAS), Bangalore - 560 065, Karnataka
14. Head, Department of Agricultural Engineering, Mahatma Phule Krishi Vidyapeeth (MPKV), Ahmednagar 413722, Maharashtra
15. Head, Department of farm Machinery and Power Engineering, College of Technology, G.B Pant University for Agriculture and Technology (GBPUAT), Pantnagar 263 145, UP
16. Head, Department of Farm Machinery and Power, College of Agricultural Engineering, Acharya NG Ranga Agricultural University (ANGRAU), Bapatla 522 101, Guntur District, AP
17. Head, Department of Farm Power and Machinery, Punjab Agricultural University (PAU), Ludhiana 141 004, Punjab
18. Head, Department of Farm Power and Machinery, College of Agricultural Engineering and Technology, Chaudhary Charan Singh Haryana Agricultural University (CCSHAU), Hissar 125 004, Haryana
19. Head, Division of Agricultural Engineering, Indian Institute of Horticultural Research (IIHR), Hessaraghatta Lake post, Bangalore 560 089, Karnataka
20. Head, Division of Agricultural Engineering, Indian Institute of Sugarcane Research (IISR), Raebareli Road, P.O. Dilkusha, Lucknow - 226 002, UP
21. Head, Farm Machinery and Post Harvest Technologies, Indian Grassland and Fodder Research Institute (IGFRI), Near Pahuj Dam, Jhansi-Gwalior Road, Jhansi 284 003, UP
22. Manager, Implements division, Tractors and Farm Equipment Limited (TAFE), 77, Nungambakkam High Road, Chennai - 600 034, TN
23. Organiser, Agricultural Tools Research Centre (ATRC), Suruchi Campus P.O. Box 4, Bardoli 394 601, Gujarat

Appendix-III

Testing centres for agricultural machineries/equipment

List of institutions approved by the Department of Agriculture and Co-operation, Ministry of Agriculture, GOI, for testing and certifying agricultural machineries and equipment

S. No.	Name of the institute
1.	Acharya NG Ranga Agriculture University (ANGRAU), Rajendra Nagar, Hyderabad, Andhra Pradesh
2.	Faculty of Agricultural Engineering, Rajendra Agriculture University, PUSA, Bihar
3.	College of Agricultural Engineering and Technology, Junagarh Agricultural University, Junagarh, Gujarat
4.	Sher-e-Kashmir University of Agri. Science and Technology, Srinagar region (J&K) and Jammu region (J&K)
5.	College of Agricultural Engineering and Technology, Choudhary Charan Singh Agriculture University, Hisar, Haryana
6.	Birsa Agriculture University, Kanke, Ranchi, Jharkhand
7.	University of Agricultural Sciences, Gandhi Krishi Vigyan Kendra, Bangalore, Karnataka
8.	College of Agricultural Engineering, UAS, Raichur, Karnataka
9.	College of Agricultural Engineering, Kerala Agricultural University, Directorate of Research, KAU (PO), Vellanikkara, Thrissur, Kerala
10.	Central Institute of Agricultural Engineering, Bhopal, Madhya Pradesh
11.	College of Agricultural Engineering and Technology, Dr. Balasaheb Sawant Konkan Krishi Vidyapeeth, Dapoli, Maharashtra
12.	College of Agricultural Engineering and Technology, Marathwada Agricultural University, Parbhani, Maharashtra
13.	Division of Agricultural Engineering, Indian Agricultural Research Institute, New Delhi

14. College of Agricultural Engineering and Technology, Orissa University of Agriculture and Technology, Bhubaneswar, Orissa
15. College of Agricultural Engineering and Technology, Punjab Agricultural University, Ludhiana, Punjab
16. College of Agricultural Engineering and Technology, Maharana Pratap University of Agriculture and Technology, Udaipur, Rajasthan
17. College of Agricultural Engineering and post harvest technology, Ranipool, Gangtok, Sikkim
18. Tamil Nadu Agricultural University, Coimbatore, Tamil Nadu
19. Sam Higginbotham Institute of Agriculture, Technology and Science, Deemed University, Allahabad, UP
20. College of Technology, Gobind Ballabh Pant University of Agriculture and Technology, Pantnagar, Uttaranchal
21. Department of Agriculture and Food Engineering, Indian Institute of Technology, Kharagpur, West Bengal
22. State level Agriculture implement testing centre, Directorate of Agricultural Engineering, Agriculture Department, Government of Chattisgarh, Teli Bandha, Gorav Path, Raipur, Chattisgarh
23. Jharkhand Agriculture Machinery Testing and training centre, Department of Agriculture and Cane Development, Directorate of soil conservation, Jharkhand, Ranchi, Jharkhand
24. State farm Machinery training cum testing institute, Faculty of Agricultural Engineering, Bidhan Chandra Krishi Viswavidyalaya, Mohanpur, Nadia, West Bengal
25. State level farm Machinery Training and Testing institute, Govt of UP, Rehmankhera, Lucknow, UP
26. State level farm Machinery Training and Testing centre, Agriculture Department, Government of Odisha, Bhubaneswar, Odisha
27. Farm Machinery Testing, Training and Production centre under Department of Farm Power and Machinery, Dr. PDKV, Akola, Maharashtra.

References

Anonymous. 1999. Crop production guide. Tamil Nadu Agricultural University and Directorate of Agriculture, Chennai. p.300

Anonymous. 2002. Agricultural statistics at a glance. Directorate of Economics and Statistics, Department of Agriculture and Cooperation, Ministry of Agriculture, Government of India.

Bhardwaj, K. C., Pandey, M.M. and Singh, G. 2004. Horticultural tools and equipments. Central Institute of Agricultural Engineering, Bhopal. p.226

Biswas, H.S. 1979. Weeding tools and implements of India. Central Institute of Agricultural Engineering, Bhopal

Chapman S.R. and Carter L.P. 1976. Crop production: Principles and Practices. WH Freeman. p 566

Das, F.C. 1981. Improved threshers developed in India. Central Institute of Agricultural Engineering, Bhopal.

Datt, S. and Sundharam, K.P.M. 1998. Indian economy. S. Chand and Co. Ltd.

Devnani, R.S. 1980. Harvesting equipment developed in India. Central Institute of Agricultural Engineering, Bhopal.

Garg, I.K. and Singh, S. 2002. Farm equipment for Punjab agriculture. Punjab Agricultural University, Ludhiana. p. 187.

Gite, L.P. and Patra, S.K. 1981. Sowing and fertilizer application equipment of India. Central Institute of Agricultural Engineering, Bhopal.

Jain, S. C. and Rai C. R. 1984. Farm tractors - Maintenance and repair.

Joginder Singh. 2006. Scope, progress and constraints of farm mechanization in India. In: Status of farm mechanization in India, IARI, March 2006

Narayanan, V. V. 1992. More on fault diagnosis in engine. *Farm Progress*, **30** (1): 8-9.

Narayanan, V. V. 1992. More on fault diagnosis in engine. *Farm Progress*, **30** (2): 7-8.

Narayanan, V. V. 1994. Braking system. *Farm Progress*, **32** (2): 8.

Narayanan, V. V. 1994. Sealed brake system. *Farm Progress*, **32** (3): 6-7.

Narayanan, V. V. 1995. Liquid ballasting. *Farm Progress*, **33** (3 and 4): 6-7.

Narayanan, V. V. 1995. Rear Wheels. *Farm Progress*, **33** (1): 7.

Narayanan, V. V. 1996. Fault Finding Chart – Rear Wheels. *Farm Progress*, **34** (1): 10.

Narayanan, V. V. 1996. Hydraulics System. *Farm Progress*, **34** (4): 9-10.

Narayanan, V. V. 1997. Hydraulics System. *Farm Progress*, **35** (2): 9-10.

Pandey, M.M. 1981. Harvesting Hand Tools of India. Central Institute of Agricultural Engineering, Bhopal.

Pandey, M.M. 1983. Power Tillers and Matching Implements. Central Institute of Agricultural Engineering, Bhopal.

Saxena, A.C., Singh, D. and Biswas, H.S. 2005. Product catalogue 2006. Central Institute of Agricultural Engineering, Bhopal.

Singh and Gajendra. 1997. Use of Energy in Agriculture and Future needs. Proceedings of the paper published in the Third Agricultural Science Congress, PAU, Ludhiana

Singh, G., Ali, N., Sahay, K.M., Dubey, A.K., Garg, V and Singh, P. L. 2000. Two decades of Agricultural Engineering research at CIAE (1978-1998). Central Institute of Agricultural Engineering, Bhopal. p. 438

Varshney, A.C., Narang, S. and Alam, A. 1995. Power tiller research and industry in India. Central Institute of Agricultural Engineering, Bhopal. p. 161.

Yadav, A., Malik, R.K., Bansal, N.K., Gupta, R.K., Singh, S. and Hobbs, P.R. 2002. Manual for using zero-till seed-cum-fertilizer drill and zero-till drill-cum-bed planter. In: Rice-Wheat Consortium Technical Bulletin Series 4, New Delhi, India. p. 27.

Colour Plates

Section 1
Tractors

Chapter 2: Engine and Associated Systems

Fig 2.3. S series Simpson Engine - Cross section

1-Web crankshaft 2-Oil gallery 3-Auxiliary drive 4-Timing case 5-Connecting rod 6- Piston 7- Cylinder block 8- Ring groove 9- Exhaust valve 10- Cylinder head 11- Valve guide 12- Valve spring 13- Rocker arm 14- Inlet valve

Chapter 3: Clutch

Fig 3.1.1. Split torque clutch

Source for fig: www.rapid-racer.com

Section 2
Agricultural Machinery

Chapter 2: Land Preparation

Two furrow MB plough

Three furrow MB plough

Fig 2.1: Mould board plough - TAFE

Two bottom disc plough

Three bottom disc plough

Fig. 2.2: Disc plough - TAFE

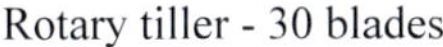

Rotary tiller - 30 blades

Rotary tiller - 48 blades

Rotary tiller - 54 blades

Fig 2.3: Variants of rotary tiller - TAFE

Fig 2.4: Power harrow - TAFE

Chapter 3: Sowing and Planting Equipment

Fig 3.1. Direct rice seeder – TNAU

Fig 3.2. Animal drawn inclined plate planter – CIAE

Fig 3.3. Air assisted seed drill – TNAU

Fig 3.4. Ridger type sugarcane cutter planter – IISR

Fig 3.5. Raised bed former cum seedling transplanter – IIHR

Fig 3.6. Potato planter - TAFE

Chapter 4: Weeding and Intercultivation

Fig 4.1. Wheel hoe – CIAE

Chapter 5: Spraying Equipment

Fig 5.1. Orchard sprayer – CIAE

Chapter 6: Harvest

Fig 6.1. Improved shear

Fig 6.2. Tractor mounted combine – TAFE

Fig 6.3. Groundnut harvester – TNAU

Fig 6.4. Potato digger – TAFE

Fig 6.5. Selective tea harvester - Williames Tea Pty Ltd

Subject Index

Note: The letter 'f' following the page numbers refer to figures.

D

E

F

V

W

Z

NIPA Publications on Agriculture Engineering

S.No.	Title	Author	ISBN	Year
1	A Handbook of Minerals,Crystals,Rocks and Ores	Alexander, P.O.	9788190723787	2009
1	A Handbook for Food Techno's	Poshadari, A.	9789381450895	2013
2	A Handbook on Irrigation and Drainage	Panigrahi, Balram	9789381450888	2013
3	Advances in Preservation and Processing Technologies of Fruits and Vegetables	Rajarathnam, S.	9789380235523	2011
4	Advances in Protected Cultivation	Singh, Brahma, Balraj Singh	9789383305179	2014
5	Agri-Food Crops: Processing,Value Addition, Packaging and Storage	R.Sasi Kumar	9789381450406	2012
6	Cereal Grains: Evaluation,Value Addition and Quality Management	Saxena, D.C.:Ed.	9789381450857	2013
7	Cereals: Processing and Nutritional Quality	Ram, Sewa and B.Mishra	9789380235073	2010
8	Computers in Agriculture	M. Kumar Shama et al.	9789385516160	2017
9	Dairy Plant Management	Puranik, D.B.	9789381450864	2014
10	Dairy Technology: Set of 2 Vols. (Set Price)	Singh, Shivashraya	9789381450994	2014
11	Dairy Technology: Vol 01: Milk and Milk Processing	Singh, Shivashraya	9789383305087	2014
12	Dairy Technology: Vol 02: Dairy Products and Quality Assurance	Singh, Shivashraya	9789383305094	2014
13	Drip and Sprinkler Irrigation	Biswas,Ranajit Kumar	9789383305766	2015
14	Drying Technologies for Foods: Fundamentals and Applications	Prabhat K. Nema, Arun S.Mujumdar	9789383305841	2015
15	Drying Technologies for Foods: Fundamentals and Applications: Part-II	Prabhat K. Nema	9789385516399	2016
16	Elements of Food Science	Jan, Safia	9789381450246	2013
17	Engineering and General Geology	Sawant, P.T.	9789380235516	2011
18	Engineering Hydrology	Pannigrahi, B.	9789385516467	2016
19	Essentials of Hydrogeology	Gurugnanam, B.	9788190851299	2009
20	Farm Machinery and Power	Powar & Aware	9788189422585	2007
21	Farm Tools and Equipment for Agriculture	Singh, Surendra	9789385516221	2016
22	Food Engineering and Technology	Sharma, H.K. & Ashutosh Upadhyay	9789383305483	2015
23	Food Process Engineering and Technology	Pare, Aakash & Mandhyan	9789380235431	2011
24	Food Processing Waste Management: Treatment and Utilization Technology	Joshi V.K. & Satish Sharma	9789380235592	2011
25	Food Science	Bawa, A.S. & O.P.Chauhan	9789381450147	2013
26	Food Science and Technology: Glossary of Preeminence	Dev Raj	9789380235806	2011

S.No.	Title	Author	ISBN	Year
27	Functional Foods	Mishra, H.N. Rajesh Kapur	9789383305988	2016
28	Functional Foods and Nutraceuticals: Sources and their Developmental Techniques	Riar, C.S. & Saxena, D.C. Ed.	9789383305964	2015
29	Geographic Information System	Gurugnanam, B.	9788190851282	2009
30	Geoinformatics Applications in Agriculture	Singh, Anil Kumar & U.K.Chopra	9788189422233	2008
31	Geomatics in Applied Geomorphology	Ramasamy, SM	9789385516429	2016
32	Geospatial Technologies for Natural Resources Mgt.	Soam,S.K. & P.D. Sreekant	9789381450802	2013
33	GIS: Fundamentals,Applications and Implementations	Elangovan, K.	9788189422165	2006
34	Hydrogeomorphology: Fundamentals,Applications and Techniques	Babar, Md.	9788189422011	2005
35	Irrigation and Agricultural Drainage Engineering	Biswas, R. Kumar	9789383305247	2015
36	Irrigation Systems Engineering	Panigrahi, Balram	9789380235387	2011
37	Mechanization of Cultivated Crops	Singh, Surendra	9789383305759	2015
38	Microbes: A Source of Energy for 21st Century	Soni, S.K.	9788189422141	2007
39	Numericals and Short Questions in Farm Machinery, Power and Energy in Agriculture	Yadav, Rajvir	9788190723718	2009
40	Postharvest Management and Processing of Fruits and Vegetables: Instant Notes	Sharma, Satish	9789380235202	2010
41	Postharvest Techniques and Management for Dry Flowers	V.Ponnuswami & Aruna,P.	9789380235868	2011
42	Postharvest Technologies for Commercial Floriculture	Verma, Anil	9789381450048	2012
43	Postharvest Technology and Engineering: An Illustrated Guide	Dev Raj	9789381450451	2012
44	Postharvest Technology and Processing of Horticultural Crops	Pandit, P.S.	9789383305278	2014
45	Postharvest Technology of Horticultural Crops: Practical Manual Series Vol 02	Sharma, Satish & M.C.Nautiyal	9788190851206	2009
46	Postharvest Technology of Horticultural Crops: Vol.07. Horticulture Science Series	Sudheer, K.P. & V.Indira	9788189422431	2007
47	Practical Isotope Hydrology	Rao, S.M.	9788189422332	2006
48	Practical Manual of Horticulture Crops: Set of 2 Vols.	Verma, Anil Kumar	9789383305711	2015
49	Practical Manual of Horticulture Crops: Vol 01: Production Technologies	Verma, Anil Kumar	9789383305704	2015
50	Practical Manual of Horticulture Crops: Vol 02: Processing and Postharvest Technologies	Vaidya, Devina	9789383305698	2015
51	Protected Cultivation of Horticultural Crops	Singh, D.K & K.V.Peter	9789383305155	2014
52	Quality Control for Value Addition in Food Processing	Dev Raj, R. Sharma & V.K.Joshi	9789380235578	2011
53	Question Bank Agricultural Engineering	Godara, Er. Amandeep	9789385516122	2016
54	Question Bank in Postharvest Technology	Guleria, SPS & Anil Kumar Verma	9789380235042	2010
55	Recycling of Industrial Effluents	Manivanan, R.	9788189422127	2006

S.No.	Title	Author	ISBN	Year
56	Remote Sensing Applications in Dryland Natural Resource Management	Gaur, Mahesh	9789381450321	2013
57	Remote Sensing in Geomorphology	Ramasamy, SM: ed.	9788189422059	2005
58	Renewable Energy Sources for Sustainable Development	Rathore, N.S.	9788189422721	2007
59	The Basics of Human Civilization: Food, Agriculture and Humanity Vol.01 Present Scenario	Prem Nath	9789381450734	2013
60	The Basics of Human Civilization: Food, Agriculture and Humanity Vol.02 Food	Prem Nath	9789383305377	2014
61	Town Planning Regeneration of Cities	Ashutosh Joshi	9788189422820	2008